Daniel Webster Cathell

**Book on the Physician Himself and Things that Concern his Reputation and Success**

Daniel Webster Cathell

**Book on the Physician Himself and Things that Concern his Reputation and Success**

ISBN/EAN: 9783337780043

Printed in Europe, USA, Canada, Australia, Japan

Cover: Foto ©berggeist007 / pixelio.de

More available books at **www.hansebooks.com**

# BOOK
## ON
# The Physician Himself

AND THINGS THAT CONCERN

## His Reputation and Success.

BY

### D. W. CATHELL, M.D.,
BALTIMORE, MD.

---

"I'd sketch the world exactly as it goes."

---

## TENTH EDITION.
### Carefully Revised and Greatly Enlarged.
(AUTHOR'S FINAL REVISION.)

PHILADELPHIA, NEW YORK, CHICAGO:
### THE F. A. DAVIS COMPANY, PUBLISHERS.
1900.

TO THE

## YOUNGER MEMBERS OF OUR PROFESSION,

AND

## ALSO TO THE OLDER ONES

Who Have Paused at Less than the Average Degree of Success in Life,

THIS LITTLE BOOK IS

## CORDIALLY DEDICATED,

With the Hope that All Who Study its Pages May be Benefited by it.

---

"Reject it not, although it bring
Appearances of some fantastic thing
At first unfolding!"

# PREFACE TO THE TENTH EDITION.

IMPRESSED with the belief that a "Book on the Physician Himself and Things that Concern His Reputation and Success" would be of decided benefit to numerous members of the profession, and finding no such work extant, I, with diffidence, attempted the duty of writing one. This book is the result.

The marked favor with which it has been received by the medical press, the expressions of approval by many well-known members of the profession, and the demand for edition after edition of it, are taken as proof that such a work was greatly needed, and that it is finding its way into the hands of many of those for whom it was written.

Grateful for this kind reception, and desiring to render it more worthy of the flattering commendations it has received, I have very carefully revised this, the tenth edition, and have also added a great deal of new material that greater experience and further reflection have dictated.

I beg you to judge it, good reader, not by opening it here or there, nor by glancing at detached paragraphs; but read it through, from cover to cover, or, better still, *study its pages*, and thus qualify yourself to weigh correctly its teachings, which I would fain have to harmonize with the advice given by the Bishop of Lonsdale to those who came to him inquiring the way to heaven: "Turn to the right, then go straight forward."

D. W. C.

1808 N. CHARLES STREET,
BALTIMORE, MD.

# BOOK

ON

# THE PHYSICIAN HIMSELF

AND THINGS THAT CONCERN

## HIS REPUTATION AND SUCCESS.

### CHAPTER I.

*"These are my thoughts; I might have spun them out into a greater length; but I think a little plot of ground thick sown is better than a great field which, for the most part of it, lies fallow."*

To FIGHT the battles of life successfully, it is as necessary for even the most skillful physician to possess a certain amount of professional tact and business sagacity as it is for a ship to have a rudder. There are gentlemen in the ranks of our profession who are perfectly acquainted with the scientific aspects of medicine, and can tell you what to do for almost every ailment that afflicts humanity, yet, nevertheless, after earnest trial, have failed to achieve reputation or acquire practice simply because they lack the *professional tact and business sagacity* that would make their other qualities successful; and there is nothing more pitiful than to see a worthy physician deficient in these respects, waiting year after year for practice, and a consequent sphere of professional usefulness, that never come.

Were any such graduate to ask me: "How can I conduct myself in the profession, and what honorable and legitimate means shall I add to my scientific knowledge and book-learning, in order to make my success in the great professional struggle more certain, more rapid, and more complete?" I should offer him the following suggestions:—

First, last, and in the midst of all, you should, as a man

and as a physician, found your expectations of success on your personal and scientific qualifications, and keep whatever is honest, whatever is true, whatever is just, and whatever is pure, foremost in your mind, and be governed by it. If you do not you will not deserve to succeed in the honorable profession of medicine, and no honest man can wish you success.

Whether, after graduation, you commence to practice without any intermediate course, or wisely strive to further prepare and refine and broaden yourself for your life's work by a limited term of service as resident physician or assistant in some hospital, or by taking a systematic course in diagnosing, prescribing, and manipulations at some post-graduate school in one of our own great cities, or endeavor to obtain a complete scientific knowledge of the profession by making a journey to the hospitals and clinics of London, Paris, Berlin, Vienna, or Leipzig, that your eyes may see the work, and that your ears may drink in the thoughts of their great teachers, is a matter of taste, money, time, and opportunity; but, whenever and wherever you commence your private practice, you should, above all else, be seriously in earnest and strive to start right, and, by the aid of hope, tireless industry, and determination, to enter promptly on the road to success; for, unless you gain popular favor by a worthy display of ability, acquire some reputation, and build up a fair practice in your first six or eight years, the probabilities are that you never will. In the battle of life it is not simply the events of school days and college hours, but the after-performances, that prove the physician.

> "Life is a sheet of paper, white,
> Whereon each man of us may write
> His word or two, and then comes night."

Beware of entangling alliances. It is, as a rule, better not to enter into partnership with other physicians. Partners are usually not equally matched in industry, capacity for work, tact, temperament, and other qualities indispensable to an intimate and congenial fellowship, and are not equally cared for by the public. Hence such professional alliances do not generally prove as beneficial or as satisfactory as expected, and consequently partnerships rarely continue long. Above all else, never ally yourself with any other physician as assistant or junior partner, to do the drudgery, or on any other terms than as an equal. The sooner you learn to depend wholly on yourself, the better. Julius Cæsar said: "I had rather be the first man in a village than the second man in a great city."

> "The fame that a man wins himself is best;
> *That,* he may call his own."

Unless you have the locality and your place of residence already selected, you may find it the most difficult problem of your life, with the whole boundless continent before you, to accurately balance and weigh the difficulties and advantages of this, that, or the other nook, corner, or opening. Whether to locate in your own town, among the generation you have grown up with, and where everybody knows all about you and your pedigree from the cradle up, or elsewhere, among strangers; in a populous city or moderate-sized town, or in a village or a rural district; in the East or the West, the North or the South of our wide-spread land, is truly a puzzling puzzle, and may be the turning-point in your life.

Many big blunders are made at the outset by locating in the wrong place; therefore, give the subject your very best thought, and decide with great care, and only after duly considering your own qualities and qualifications, as well as the locations, —whether you are self-reliant and pushing, or quiet and unobtrusive; whether you have abilities that will enable you to compete with the wisest and the best, and compel people in a populous centre to employ you in preference to your neighbors; or whether, being less fully armed, you had better be satisfied with mediocrity, and become a modest country doctor in a less thickly settled location, where there is less competition and less talent to encounter, or go to the new States and grow with the growth of the settlements, or rise with the villages, or spread with the cities that are springing up.  I may remind you, however, that

> "Where there is nothing great to be done, a great man is impossible."

Medicine, like everything else, thrives best in good ground. By all means seek to locate in a community to which you are suited; that will be congenial as a place to live in, and in which you are likely to get business and be useful to your fellow-beings, and also to earn a living for yourself.  Bear in mind that unpopular opinions in politics or religion injure, and that, all else equal, you will be more likely to succeed and be contented in a section where your views, habits, and tastes are naturally in harmony with the bulk of the people, morally, socially, and politically.

No matter where you start, if, alas!

> "You wear the bloom of youth upon your cheek,"

you will hear the adjective "young" oftener than is pleasant, and encounter up-hill difficulties that older physicians do not. "He looks too young;" "He lacks experience;" "He don't know anything;" "He has no practice, therefore is no good;" "He shouldn't doctor me," and "I'd send him off and get an older physician," are among the often-heard expressions. Face them all bravely. Never doubt, but show the world, by good management and good habits, that you deserve to succeed, and success will surely come. Strict attention to the opportunities that will present themselves for winning confidence in cases that are incidentally thrown into your hands, and a diligent cultivation of your talents, with promptness, civility, courtesy, and unobjectionable conduct to all, rich and poor, and pleasant manners, but no time to gossip, will bring it. Even a single event, or an accident, may fortunately give you an introduction to extensive business.

If you are young and youthful-looking, unless you have some special reason to the contrary, let your beard grow, if it will; and remember that our Saviour and Alexander each lived but thirty-three years, and Napoleon commanded the army of Italy at twenty-seven.

If you begin practice in a city or town, the location and appearance of your office will, more or less, affect your progress, and you will do well to select one, easy of access, in a genteel neighborhood, upon or very near one or more of the main thoroughfares, and convenient to either a densely populated old section, or a rapidly growing new one. The nearer to busy centres of mechanics and laborers, the better. If you were to locate on a back or unfrequented street or other out-of-the-way place, or in the country, where the land is unproductive and the population sparse, simply because there is but little or no competition, it would naturally suggest to the public that you had poor judgment or were made of timorous, negative material, or lacked the spirit of enterprise and enthusiasm, or were waiting for practice to come naturally, and for success to be handed to you on a silver platter, or else had defective ambition and distrust of your own acquirements.

"He that does not show himself is overlooked."

Remember, in making your selection, that a physician cannot rely on his near neighbors for patronage; people in your immediate neighborhood may never employ you, while some farther away will have no one else.

If your first location disappoints you, remove to another; but avoid frequent removals, and do not shift or change from one place to another unless it is clearly to better yourself. Select a place suited to your abilities and taste, and then be tenacious. Reputation is a thing that grows slowly, and every distant removal imperils one's practice, necessitates new labor, and sometimes even compels a commencing of life over again. A physician's frequent removals may also create a bad impression, and look like natural instability or dissatisfaction from lack of success.

Branch offices are, as a rule, not desirable, for, in addition to the loss of time, and wear and tear in going to and fro, and double trouble in general, they are apt to create an uncertainty in the minds of those who may be in want of the physician, as to where and when he is most likely to be found. On estimating all the advantages and disadvantages, it will probably be found that, as a rule, a plurality of offices increases greatly neither one's practice, one's popularity, nor one's income, but does add greatly to the labor, and hence may be regarded as likely to prove more annoying than profitable.

It has been said that

> "A physician never gets bread
> Till he has no teeth to eat it."

Be this as it may, it is risky for you, if a beginner, with no influence and but little money, to locate in a section already overstocked with popular, energetic physicians, as their superior advantages, established reputations, and warm competition may keep you limited and crippled for too long a time before a chance or a change comes. Also, guard against going too close to large, free hospitals and dispensaries. Your first necessity is to possess knowledge and skill as a physician, your second is to find a field in which to exercise and display them; but, no matter where you locate, if you expect business immediately to follow your annunciation of being ready to receive it, you will, except under very extraordinary circumstances, be rudely disappointed.

A corner house is naturally preferable to one in the middle of a row, since it is convenient for persons coming from all directions, and not only has facilities for constructing an office entrance on the side street, leaving the front door free for other callers, but also insures to the consulting-room a good light for examinations, operations, and study.

Regarding offices: Try to have a good, comfortable waiting-room, with a recessed front door; also, a good, light, airy, and accessible consulting-room of moderate dimensions, with, if at all convenient, two doors,—one for the entrance and the other for the exit of patients,—for many of those who consult you will prefer to be let out through a passage or private door, and thus escape the gaze and possibly the comments of others in waiting.

Exercise special care in their arrangement; give them a pleasing exterior and make them look fresh, neat, and clean outside; and inside, give them a snug, bright, and cosy medical tone, and let their essential features show that their occupant is possessed of good taste and gentility, as well as learning and skill; and that they are not a lawyer's consulting-rooms, nor a clergyman's sanctum, nor an instrument-maker's shop, nor a smoking-club's head-quarters, nor a sportsman's rest, nor a loafing room for the gay, the idle, the dissipated, and the unemployed; nor a family parlor, nor a social meeting-place of any kind; but the offices of an earnest, working, scientific physician, who has a library, takes the journals, and makes full use of the instruments and methods that science has devised for him, and regards his office as the twin sister to the sick-room.

Take particular care, however, to avoid making a quackish display of instruments and tools, and keep from sight such inappropriate and repulsive objects as catheters, syringes, stomach-pumps, obstetric forceps, splints, trusses, amputating knives, skeletons, grinning skulls, jars of amputated extremities, tumors, manikins, the unripe fruit of the uterus, etc. Also, avoid such chilling or coarse habits as keeping vaginal specula or human bones on your desk for paper-weights.

But while you should make no undue exhibition of books, surgical instruments, etc., it is not unprofessional to have about you—not for display, but for ready and actual use—your outfit: microscope, stethoscope, laryngoscope, ophthalmoscope, spirit-lamp, test-tubes, reagents for testing urine, and the various other aids to precision in diagnosis, and the numerous instruments you make use of in treatment; also, to ornament your office with diplomas, certificates of society membership, potted or cut flowers or growing plants or vines, fine etchings, pictures of eminent professional friends or teachers, or of medical celebrities,—Hippocrates, Galen, Harvey, Gross, or whomever else you specially admire; academical prizes, professional relics, keepsakes, mementoes, medals, or anything else that tells of your mental or physical prowess in earlier days, or is specially associated with your medical studies and career. But, unless it be a few artistic ornaments or works of art, it is better to limit such articles to those having relation to you as a student or physician.

In buying your office outfit see that the walls and floors are tastefully covered. Articles of furniture should be few in number, but good, including a small, and if means will admit, handsome book-case, with writing-table and chairs to correspond. Have comfortable chairs for your patients' use, so arranged that they may sit in a good light during examination, but beware of stocking yourself with novelties and instruments that will probably go out of fashion or rust or spoil before you will need them. It is prudent not to invest heavily at first, and to wait and buy none but the usual every-day instruments, which the urgency of certain cases will not give you time to go for, when occasion arises for their use, until you have a use for others. Bear in mind that soft-rubber goods, and soft goods generally, deteriorate and finally become worthless in keeping.

A neat case of well-labeled and well-corked medicines, or a cabinet of minerals, is of use and not unornamental; so also are dictionaries, encyclopædias, and lexicons for ready reference; also, a non-striking time-piece to quietly notify the time to physician and patient by its tick-tick-tick; but display no miniature museum of sharks' heads, stuffed alligators, tortoise-shells, impaled butterflies, bugs, ships, steam-boats, mummies, snakes, fossils, stuffed birds, lizards, crocodiles, beetles, tape-worms, devil-fish, ostrich-eggs, hornets' nests, or anything else that will advertise you in any other light than that of a physician. It will, to the thinking portion of the public, seem very much more appropriate for you, as a physician, to be jubilant over a restored patient or a useful medical discovery than to be ecstatic over a stuffed flying-fish, an Egyptian mummy, or a rare shell.

If you have a natural love for such incongruous things, or are a bird- or dog- fancier, or a bug-hunter, at least keep the fact private and keep your specimens out of sight of the public, and endeavor to lead every one to think of you only as a physician.

It is your duty, as well as your interest, to display no political or religious emblems, portraits, etc., about your office, because these relate to your personal sentiments; being emphatically a public man, and your office a public place, not for any special class, but for every faith and party, no matter what shade of partisan or sectarian pictures you may display, they will surely be repugnant to some,—

"On life's stormy ocean diversely we sail,"—

and in this and other matters fairly open to criticism it is a wise maxim to respect public opinion. Difference in religion or politics has often either prevented the employment of physicians or caused their dismissal, and the obtrusion of unpopular political or religious views has marred the prospects of many a physician; besides, what is popular to-day may be unpopular to-morrow; therefore, keep your heart and your office open to all denominations and to all parties. This will recommend you equally to all.

Establish a regular professional and business policy at the beginning of your career. Be at your post as punctually as possible, and have your office lighted regularly every evening at the proper hour, your door-bell answered promptly, professional messages entered on the slate by the person in charge, and in all other respects show punctuality and system. You will find that absence from your office when needed, particularly if away for sport or pleasure, is a fruitful source of loss of practice; if, on the contrary, you are at your post, people will credit you with seriousness in your profession, which will advertise you and bring you patronage.

Do not allow the ladies of the family to lounge about your office, or read your books, answer the office-bell, etc., lest it repel certain kinds of desirable patients. Both messengers and patients would rather meet you or your servant than ladies.

You should respect public opinion in this and in all other matters justly open to criticism.

Still more important to success will be the morals of the companions you make in your early career; in fact, all through life a physician is judged by the company he keeps. Avoid associating with aimless idlers and those who bear a merited stigma, or are notoriously deficient, or whose hopes and ambitions have been blighted or wrecked by intemperance; or their good names otherwise tarnished by their own misconduct. On the contrary, let your associations be, as far as possible, with professional brethren and people of genuine worth. Prefer to spend your unoccupied moments in your office with your standard works and medical journals, or in rational conversation with high-minded friends, or other physicians, or at medical meetings, or at the medical library, to lounging around drug-stores, hotel-bars, saloons, club-rooms, cigar-stores, billiard-parlors, barber-shops, or corner-groceries, with lazy fellows, who love doing nothing, frivolity, and dissipation; or to taking such persons out riding in your carriage, or to the horse-races, or to join the throng at the base-ball game. No ordinary man ever conceives a more exalted opinion of a professional man by fraternizing with him at such places, or in seeing him in such company.

As a further but minor aid to successful progress, be courteous to all kinds of patients with whom necessity or duty brings you into contact; but while you treat all men as brothers, and all women as sisters, beware of talking too freely, and do not handshake and harmonize and associate with the coarse, ignorant, and unappreciative indiscriminately, for undue familiarity shears many juniors of influence and prestige. Also, never become so familiar as to lay formality aside and enter a patient's house or room without announcing yourself by a gentle rap or ring at the door.

Avoid companionship with quacks and irregulars, as it would detract from both you and rational medicine which you represent and give countenance to delusions and pretenders. Shun this and every other contaminating alliance that would confound them with us before the public.

What shall be said regarding self-mutilation with harlots and association with varnished concubines? Of drinking and of gambling? Of the dethroning fields of Venus and Bacchus! Oh! physician, if you have entered either of these DANGEROUS roads, follow the dictates of common sense, and turn from it this day, this hour! for they both lead rapidly downward, and either of them will deform and warp all your finer sensibilities, prove fatal to every ambition, and speedily put a death-blight on all your prospects. And if indulging any one of these habits singly will be like sowing dragon's teeth for yourself, what will be the combined effects of them all? It will insure social and moral death! Professional suicide,—short, quick, and sure! while your relatives and friends will weep in all the bitterness of disappointed hope for your dishonorable downfall.

"Too late to grieve when the chance is past."

An unspotted, honorable name is the only thing that will render your life happy and enable you successfully to withstand the critics, for neither you nor any other physician can successfully lead a double life, or afford to despise public opinion.

"A pebble in the streamlet scant
Has turn'd the course of many a river."

Unfortunate acquaintances have been the downfall and ruin of many a promising young physician; therefore, select your associates with great care, and do not let your office be a lounging place or a smoking-room for horse-jockeys, dog-fanciers, base-ballers, politicians, chatty blockheads, or others whose time hangs heavily on their hands. The public look upon physicians as public characters,—earnest, sober, studious men, with scientific tastes and temperate habits, who have been singled out and set apart for a lofty purpose, and as socially, mentally, and morally worthy of an esteem not accorded to such people, or even to ordinary citizens engaged in the private business of life. The idle jokes, childish amusements, boyish gambols, commonplace gabble, and tone of thought common to light-minded people do not harmonize with the studies, tastes, and desires of worthy physicians, and, moreover, tend to weaken or destroy the faith of the public, which is so essential in our work, for on no profession does faith have such influence as on ours. You as a physician are public property, and the public, and especially the female portion of it, with eyes like a microscope, will take cognizance of your associations and of a thousand other little facts regarding you.

"Things small in themselves have often a far-reaching significance."

In fact, every circumstance in your appearance—dress, manners, actions, walk, speech, conversation, habits, where you are to be found when not professionally engaged, etc.—will be closely observed and criticised in order to arrive at a true verdict, more especially in the early years of your career. The question will never be asked whether you were graduated at the new or the old college, or whether from the "college of wigs, or abroad," but it will be, "Is he a good physician?"

Put not a feather's weight upon the honorable ambition of any one, or a straw in the pathway of his worthy aspirations, but be very cautious how you involve yourself by inducing persons to study medicine, as there are already three physicians where one is required. Besides, their failure in the profession, or their misconduct, or their unfair rivalry may, in time to come, work great injury to you.

"Out of a white egg often comes a black chicken."

Besides, it is neither profitable nor advisable for you, a private practitioner, to take aspirants for Æsculapian honors as office students, as they will necessarily be in the way and divert your mind from other duties; but, if you do take any, charge them a fair price for the privilege, and remember that in taking students you stand as a guardian at one of the outer gates of the profession, and listen only to such applicants, rich or poor, as have a pure, high-souled, and just appreciation of the profession, well-balanced, good sense, sobriety, mental and physical vigor, good habits, intellectual capacity, natural aptitude, and a strictly honorable ambition or enthusiasm to be a worthy physician.

Remember that you cannot polish a fungus or make a sponge shine, and that good gas makes a good light and bad gas a poor one; that a good battery generates good electricity, and that a bad one necessarily makes a poor kind; so, also, that a good brain, a good mental soil, creates better ideas and bears better fruit than an ordinary one. A high-thinking, practical-minded youth from the corn-field or a log-cabin, with scarcely enough clothes to hide his nakedness, and the aimless son of a millionaire may each apply. If you take either, be not long in choosing. Brains and common sense are a rare gift from heaven; and a diploma from every medical college on the face of the earth, each bedizened with ribbons—red, white, and blue—and each stowed away in a gold case set with diamonds, cannot give them to those who lack them. Bear this in mind, and dissuade and refuse every one who has been seduced from his true calling in humble life to embrace medicine, from a belief that its study is merely a pleasurable pastime, or that it is simply a trade, or that it is less laborious than the business he is following; or Jacks-at-all-trades, who are tempted to add M.D. to their list, by the ease with which a "sheepskin" can be obtained; or by the false notion that to be a physician is a gay and pleasant life, or a smooth and rosy road to money-making; or simply to please a fond grandmother, or a doting papa; or from a false dream of an easy life. Also, turn your back on the callous, the tough, and the ox-hearted, rough-fisted fellow, who boasts that he is stony, can stand anything, and wants to be a surgeon, because he feels an anxiety to see the shedding of human blood, or any other applicant so unworthy.

The popular opinion that now the untilled, thoughtless, brainless bumpkin, who has hardly mastered the multiplication table, and knows not the difference between an angle and a triangle, can stop following the plough, or driving the jack-plane, or drop the yard-stick, or pen, or teacher's rod; or desert his lap-stone and bad shoemaking to-day and in a few months be metamorphosed into an M.D.,—

"While all who know him wonder how he passed,"—

and that an ornamental sign or a fancy door-plate with a name (and the prefix Doctor) on it, with a buggy at the door, is about all that is necessary, is now causing thousands of young men to quit their proper avocations in life and study medicine, only to fail in its pursuit.

In getting your office signs or door-plates, remember that a physician has them not as advertisements, but simply to show his office to those looking for him. Your signs should be neither too large nor too numerous. One of black smalt with gold letters is the neatest and most attractive of all; one such sign on the front wall for the day-time, and a glass one with black letters in the window, to be seen at night, when your office is lighted, are sufficient. The letters on the former should be round and well shaped, and not more than two inches high, with corresponding width. A polished brass sign, engraved with your name, and the letters filled in with black, and mounted on a finished, hard-wood board, is also neat and stylish.

All signs should be neatly made and correctly lettered, for even one's sign makes an impression, either good or bad, on the public, and first impressions are very enduring.

In this country it is better to put Dr. . . . on your sign or door-plate than to put . . . , M.D. "Doctor" looks better, and is understood by all; but to speak of yourself as a physician rather than a doctor, or to refer to your professional brethren as physicians rather than doctors, sounds more distinctive and falls better on the ear.

To put "Physician and Surgeon" or "Physician and Accoucheur," or other compound addition, on your sign would seem unnecessary in this region, since all physicians (except the specialists) are supposed to be surgeons, accoucheurs, etc. The practice of medicine on the human body now allows no such This-or-That division of learning, and all are blended by the law; the medical case of to-day may be the surgical or obstetrical case of to-morrow; almost as well might the confectioner's sign say "Cold Ice-Cream."

Unless your name is likely to be confounded with that of some other physician, it will be well to omit your given name or initials from your signs or door-plate; but it should be on your cards. Of course, if your name is "Smith," or "Jones," or "Brown," it would be necessary to put your given name on your signs; but if your name is uncommon, it is not. People will not speak of Doctor John W. Garfield, but of Doctor Garfield.

Do not allow other people's signs of tooth-drawing, cupping and leeching, millinery, dressmaking, painting and glazing, boarding, etc., in company with yours.

The lettering on your window-glass may be protected from being scratched, or otherwise defaced, by having a pane of common glass placed behind the lettered one.

It is deemed unprofessional to state where you graduated and how long you have practiced, upon your cards and signs, or in the newspapers.

Adopt regular office hours early in your career, and post them conspicuously in your office; also, have them on your cards.

It may be a question whether it is advantageous to have a sign designating your office hours on your office window, or on the house front, to be seen by the outside public. Your situation in business should influence your decision on this point. A young physician, or one who has much spare time at home, in addition to his stated hours, will be more apt to catch the overflow, emergencies, cases of accident, calls from those who are strangers in the city, and other anxious seekers for "any one, so he is a physician," and who have perhaps searched and found all the busier physicians away from their offices, if an exhibition of his office hours does not drive them off by telling them before ringing the bell that they have come at the wrong time, when in fact he is at home wishing for calls. On the contrary, one busily engaged in outside practice, who has no other time for office consultations than the specified hours, can, by displaying them outside, regulate his business, and prevent various annoyances, by letting every one see his hours before ringing.

An excellent rule is to direct attention to both the beginning and ending of your office consultation hours, as: "Morning office hours begin at 7 and end at 9; afternoon office hours begin," etc. Or: "Office hours: morning, between 8 and 9 o'clock; afternoon, between 1 and 3 o'clock," etc. Many people inconsiderately think that as your office hours are from 7 to 9, if they get there one minute before 9 o'clock they are in time; whereas, if they come at that time they will be sure to keep you past your hour for beginning your outside professional work. By regulating your time thus, and constantly urging those you attend to observe your home hours strictly, you can accomplish doubly as much, with less hurry and more satisfaction to all. Indeed, by persistently schooling patients to observe these hours, and to send for you, as far as practicable, before your accustomed time for starting on your regular rounds, preferably in the morning, you will do much to systematize your business, and to lessen the number of calls at odd and inconvenient times, which do so much to increase the hardships of the physician's life. For persistent late-comers to come strolling into a busy physician's office for advice at odd or unseasonable hours, or at seasons allotted to privacy and rest, amounts almost to persecution. So, also, does having to visit the same neighborhood half a dozen times a day, in consequence of his patients not sending for him before he leaves home to commence his rounds. The time allotted to office patients may be greatly curtailed by naming certain times at which you can be found at your office;

for instance, instead of having it "Morning, from 8 to 9 o'clock; afternoon, from 1 to 3 o'clock; evening, from 6 to 8 o'clock," have the sign read, " Office hours: morning, *about* 8 o'clock; afternoon, *about* 2 o'clock; and evening, *about* 7 o'clock," which times are easily remembered, and will cause all who come to get there *about* those hours.

If you should ever get very busy, and be pressed for time, your sign might still further emphasize it, after stating your hours, by adding, " No office consultation at other hours."

Have a slate in a convenient place, whereon messages may be left during your absence from your office, and have over it a little sign something like this: " In leaving a message for the Doctor, be careful to write the name, street, and number."

You should keep a supply of cards, with your name, residence, and office hours on them. An inch and three-fourths by three inches make a good size.

It is also necessary to keep a supply of small and neat blank bills, and to have envelopes and paper with your name and address printed on them. Let your bills read, " For professional services." Blank forms for use in giving certificates to sick members of societies, etc., are also very useful. Printed professional certificates look much better and more formal, and generally give more satisfaction than written ones.

A speaking-tube, from your outside office door to your bedroom, prevents exposure to raw night-air at an open window, and is of great utility when your night-bell rings.

The telephone is also both a luxury and a necessity. Many physicians, however, are deterred from having one by the fear that it will cause them to be summoned to patients, good and bad, at a distance too great for them to attend, or that its convenience will cause annoying calls and messages to be sent at unseasonable hours. This belief is erroneous. The telephone really does the opposite, and enables one to resist the arguments and attempts at persuasion so often encountered in personal interviews. It is, moreover, far easier to decline to pay a visit, to urge a plea, to suggest a remedy, or give direct instructions through the telephone, than by an interview with a fallible messenger. If you have a telephone, put its number on your cards, bills, envelopes, letter-paper, etc.

On commencing practice, you should get a pocket visiting-list, a cash-book, and a ledger, and commence to keep regular accounts at once, taking care to "post up" regularly either weekly or monthly; this will teach you system, and in the course of time save you thousands of dollars.

Be careful to record the full name, occupation, and residence of every new patient; for, although the identity of this one and that one may, at the time, be very clear in your mind, yet as patients increase and multiply and years elapse, your personal recollection of each will become misty and confused, and consequently may entail on you considerable money loss. Method in business is one of its chief instruments. Also, never neglect to jot down memoranda of office consultations, payments, new calls, etc., in your visiting-list, with a lead-pencil, until you get an opportunity to write them in ink.

One's visiting-list can be most conveniently carried in a wide but shallow pants pocket on the left hip.

It is well to have a copy of the fee-table framed and hung in a suitable place in your office, that you may refer patients to it whenever occasion requires. It is also wise to have a small, neat sign, with "Office Consultations from $1 to $10, cash," posted in some semi-prominent place in your office. It will show your rule and tell your charge; it will also remind any who might forget to pay of the fact, and by confronting less honest people will put them in a dilemma. You can, when necessary, point any one to it and ask him for your fee; it will also give you a chance to let him know you keep no books for office patients. Such a sign will save you many a misunderstanding and many a dollar. Of course, you may omit its cash enforcement toward patients with whom you have a regular account.

Having your charge from "$1 to $10" will enable you to

get an extra fee for cases of an extraordinary character, and still allow you to charge minimum fees for ordinary cases. Such a schedule will also make those who get off by paying the lowest fees feel gratified, and will show everybody that you assume to be skillful enough to attend $10 cases.

Cultivate office-consultation practice assiduously, for it is a fertile source of reputation and of cash fees; attending such patients as are able to go out-doors, at your own office, is a great saving of time and fatigue to the physician. Strive to benefit and give satisfaction to every patient who comes to consult you, that every one may go away impressed with a belief that the nature of his malady is clearly recognized and understood, and that you will do your best to remedy it; for each will, while there, form some definite opinion in regard to you, and will ever after give you either a good or a bad name.

Keep a small case of medicines at your office representing the most frequently employed articles of the pharmacopœia, especially during the first years of practice; handling them will not only familiarize you with their appearance, odor, miscibility, taste, and other characteristics, but also enable you to get your fees from unreliable patients, and such others as can appreciate advice and something tangible combined, but who cannot properly value advice alone. Besides, by keeping cathartic pills, aromatic spirits of ammonia, lime-water, morphia granules, etc., you can save yourself many a tramp at night, during storms, on Sunday, great holidays, at odd hours, etc., by sending a suitable remedy by the messenger; and give the patient both relief and satisfaction, till you can go. You have a perfect right to supply a patient with medicine if you choose. Very extensive home dispensing, or running a rudimentary drug-store, or a pill and globule traffic, however, tends to consume time that might be much better employed, and to dwarf one in other ways. Furnishing his own medicines does not pay, if a physician is established in good, reliable circles, because it is far better for him to base his charges squarely on the abstract value of his time and skill. Besides, one's high tariff and rough compounding would engender the criticism and enmity of neighboring druggists and others. Never under any circumstances sell medicines to any but your own patients.

Dispatch every professional duty promptly and punctually, so as to get it out of the way of whatever may happen to come after. When summoned to cases of colic, convulsions, accident, etc., go, if possible, immediately. Then, if you are too late to be of service, you will neither have cause for self-reproach nor be responsible for default of duty. When you cannot go at once without neglecting another pressing case that has a prior claim on your services, or other duties equally as urgent, it is much more satisfactory to your patient if you send a remedy, with instructions for use until you can go, than to write a prescription; because, to send a prescription in such cases seems rather as if you do not sympathize, or as if the patient was on your don't-care-to-attend list, and, if the case takes an unfavorable turn, or does not eventuate favorably, you may be blamed and criticised.

> "Of all sad words of tongue or pen,
> The saddest are these: it might have been."

When you reach a patient whose friends have, in the excitement, sent for a number of physicians, with no special choice among them, it is well to have them promptly send a trusty messenger or a courteous note to the others to cancel the call and save them trouble, by informing them their services will not be required.

If, at your office and elsewhere, you make a judicious and intelligent use of your instruments of precision,—the stethoscope, ophthalmoscope, laryngoscope, the clinical thermometer, the tape, the microscope, and the reagents necessary to a careful examination of tumors, sputa, calculi, urinary disorders, etc.,—they will not only assist you very materially in diagnosis, but will also aid you greatly in curing nervous and terrified people, by increasing their confidence in your ability, and enlisting their sympathetic concurrence in your remedial treatment.

Always carry with you, in your professional rounds, a good clinical thermometer, female catheter, bistoury, hypodermatic syringe, small forceps, lunar caustic, probe, needles, pen-knife, etc., for ready use. Keep a little raw cotton in the case with your clinical thermometer to protect it against breakage, and never omit to wash it and all other instruments immediately after use.

Be especially careful to avoid syphilitic inoculation, septicæmia, etc., and never, under any circumstances, use a cut or an abraded finger in making vaginal and other examinations; if your preferable hand is unsafe, use the other. Cosmolin and vaselin answer a good purpose; they have no affinity for moisture, and both keep for years without becoming rancid or decomposing. Get a supply of either, and keep it in your office for anointing your fingers, instruments, etc. Wooden tooth-picks and cigar-lighters are also very handy for making mops, applying caustics, etc. Being inexpensive, each one can be thrown away after one service, instead of being kept for further use, as must be done with expensive articles.

Knives, probes, needles, and other instruments can be readily cleaned and disinfected, both before and after being used, by thrusting them several times through a wet, well-soaped towel or rag, or into a cake of wet soap.

You should have a special receptacle in your office for cast-off dressings from cases of gonorrhœa, syphilis, septic ulcers, and other filthy affections, which, when they accumulate, should be burned.

With the view to maintain your physical health, you should endeavor to live temperately and comfortably, and to rest as much as possible on Sundays and at night; and, moreover, if you would avoid the risk of break-down in health, as happens to hundreds of our profession, make it a cardinal point of duty to yourself and family to get your meals and sleep as regularly as possible, and to keep your digestion in order; then you need have but little fear of overwork.

A decent respect for the opinion of the world should lead

you to practice all that constitutes politeness in dress and deportment. Keep yourself neat and tidy, and avoid everything approaching carelessness or neglect. Do not altogether ignore the fashions of the day, for a due regard to the customs prevailing around you will show your good sense and discretion. Even though the prevailing style of dress or living borders on the absurd or extravagant, it may still be wise to conform to it to a certain extent. Young says:—

> "Though wrong the mode, comply; more sense is shewn
> In wearing others' follies than our own."

You never heard of a designing swindler, or a confidence-man, or a gambler, or a pseudo-gentleman of any kind, who dressed shabbily or in bad taste, for

> "These men's souls are in their clothes."

Such people are all close students of human nature, and, no matter how abandoned they are, no matter how tarnished in character or how blackened in heart, they too often manage to hide their deformities as with a veil from all but the few who know their true characters, by assuming the dress, manners, and tone of gentlemen. Now, if genteel dress, polished manners, and cultured address can do so much for such fallen specimens of mankind, how much greater influence must appearance, manners, and voice exert for those who are truly gentlemen and members of an honorable profession.

Nevertheless, do not, under any plea, be a leader or patronizer of loud or frivolous fashions, as though your starchy foppishness and love of fine clothes had overshadowed all else; discard, also, glaring neckties, flashy breastpins, loud watch-seals, brilliant rings, fancy canes, cologne, perfumes, attitudinizing, and all other peculiarities in your dress or actions that indicate overweening self-confidence, or a desire, with the assistance of the Graces, to be regarded as a man of fashion, or a swell.

> "Cupid, have mercy!"

Such peacockish individuals may be admired, but they are not usually chosen by discerning persons seeking a guardian for their health.

Even though you be ever so poor, let your garb show genteel poverty, for every physician's dress, manners, and bearing should agree with his noble and dignified calling. The neglect of neatness of dress and the want of polite, refined manners might cause you to be criticised or shunned. You will sometimes see superficial but spruce little Dr. Tact, whose head is comparatively empty, who always sat on the back benches at college, succeed in getting extensive and lucrative practice, and paying heavy bills for horseshoes, almost entirely by attention to the outer trappings and affability of manner; while Dr. Profundus, Dr. Alltrue, and Dr. Talent, professionally more able and personally more worthy, will languish, and never learn the cost of carriages and the price of horse-feed, by reason of defects in these apparently trivial matters. Alas!

"Veneering often outshines the solid wood.

Clean hands, well-shaved face or neatly-trimmed beard, unsoiled shirt and collar, an unimpeachable hat, polished boots, spotless cuffs, well-fitting gloves, fashionable clothing, cane, sun-umbrella, all relate to personal hygiene, severally indicate gentility and self-respect, and naturally impart to their possessor a pleasurable consciousness of being well dressed and presentable.

"I am not a handsome man, but my beaver doth lend me an air of respectability."

The majority of people will employ a tidy, well-dressed physician, of equal or even inferior talent, more readily than a slovenly one; they will also accord to him more confidence, and expect from and willingly pay to him larger bills.

Avoid extraneous pursuits and a multiplicity of callings, especially such as would interfere with your duties as a physician, or would give you a distaste for the profession, or cause you to resume its duties with a feeling of irksomeness. Divorce medicine from all other avocations, however important, respectable, or lucrative,—from the drug business, dealing in petroleum, or salt, or cattle, or horses; nor be equally interested in the practice of medicine and in school-teaching, or in **pushing** the jack-plane, or in following the plow; or giving public readings, or preaching, scribbling poetry on subjects not connected with medicine, or fiddling or singing at concerts; or base-ball playing, rowing matches, amateur photographing, etc.,—because the public cannot appreciate you or any one else in two dissimilar characters or in two incompatible pursuits: half physician and half druggist, or three-eighths physician and five-eighths politician, or one-third physician and two-thirds sportsman, or other similar mixture of incongruities: for it is in medicine as in religion,—no one can serve two masters. Of course, if you choose to change off and quit medicine for any other calling, it is legitimate to do so; but it is better to be a whole one thing or another.

Although it may seem paradoxical, even reputation as a surgeon (though surgery is but a branch of our art), or as a specialist of any kind, militates decidedly against reputation in other departments of medicine. The public in general believe that a surgeon, with his sharp saws and thirsty knives, delights in spilling blood, and is good only for *whipping off* limbs, or other cutting operations, and that a specialist is good only for his specialty, just as a preacher is for preaching.

Hesitate even to take such offices as vaccine physician, coroner, city-dispensary physician, sanitary inspector, etc., in a section where you expect to practice in future, more especially if you must have illiterate political demagogues or buffoons for employers or companions.

"Jack in office is a great man."

All such functions tend to dwarf one's ultimate progress, and sometimes create a low-grade reputation that it is hard to outlive. To many people, taking such offices for the fees to be obtained looks somewhat like a confession of impecuniosity or of inferiority, and creates an adverse impression that is not overcome for years. If you have any merit at all, and an open field, private practice industriously followed will lead by better roads to greater success.

The last remark is, also, to a certain extent true of the position of permanent physician or assistant physician to hospitals, infirmaries, lunatic asylums, dispensaries, almshouses, reformatory or penal institutions; or in the army, or on board emigrant or naval vessels, where employment in a snug or easy job, at a petty salary and the comforts of a home, for a few of his most precious years, have caused many a physician fully qualified for success as a practitioner to pass the flower of his days, to lose the best, the golden part of his life, and let slip opportunities that could never be recalled.

> "Too soon, too soon
> The noon will be the afternoon;
> Too soon to-day will be yesterday."

Bear in mind that such positions can never be depended on longer than those in power find it to their interest to change.

If you ever become a teacher of medicine in a college, with a choice of branch, instead of taking Physiology, Materia Medica, Jurisprudence, Hygiene, or other non-personal subjects, take care to aim for a practical chair, one that relates directly to the sick, and that is likely to increase your skill, get you special work to do, or otherwise advance your reputation and your private practice.

Shun politics and electioneering tactics; for politics, even when honorably pursued, are injurious to a young physician's prospects; later, when his medical reputation is already extensive, they generally lessen his professional popularity, although they may not necessarily ruin him. If the best of good politics injure thus, how much worse is it to be dabbling in the dirty pools of partisan politics, at ward rallies and bar-room conferences, or plunging into demagogism, and wire-pulling, slate-making, log-rolling and pipe-laying at primary meetings, caucuses, conventions, etc., with "the b'hoys." No! no! thrice no! For, besides escaping many anxious hours and bitter disappointments, you can in the long run make ten friends and ten dollars by being no man's man, and calmly sticking to your profession, while you are making one of either in the polluted and polluting waters of party politics, lending your name to help the campaign, or intriguing and scrambling for office with those who belong to the parties chiefly for their loaves and fishes.

Array yourself on the side of morality, virtue, honesty, religion, etc., but neither make your religion nor your irreligion a stepping-stone to practice, and never join a church or a religious society for the purpose of gaining popularity or church influence. You will surely find that society, church, political, and other special groups of sectarian patients, gained because they belong to the same society or party in politics, or are affiliated with you in society matters, or go to the same church, or because you deal with them in business, or live on the same street, or because they like the way you walk or dress, rather than through appreciation of your merits as a physician, are neither very profitable nor very constant. If, instead, you will banish everything that comes between you and your legitimate work, try to bring practice *by your practice*, and cultivate patients secured promiscuously from all parties, and from every direction, because they believe that you, as a physician, possess solid merit; and have faith in your brain and your heart and your hand; it will in the long run make you more friends, and firmer friends, and pay you better than attending solely to any one political sect or religious creed, or following any other outside issue.

A riding physician has several advantages over the one who makes his rounds on foot; not only is he able to see a greater number in a given time, and with much less fatigue to himself; but he gets rest while riding from one patient to another, and can spend the time in thinking; can collect and concentrate his mind more fully on his serious and puzzling cases while riding than if walking, and when he reaches his patient he is in good mental and physical condition to begin his duties, while the walking physician arrives out of breath, excited, and in need of rest. The former can prescribe and be gone while the latter is waiting to regain his breath. Another convenience is, that Tenderfoot salutes acquaintances as his carriage meets them and rides on; whereas, Trudger is compelled to stop, and loses valuable time in conversing with convalescent patients, old friends, and others.

You should, therefore, get a good-looking horse and a genteel carriage as soon as your circumstances will justify. Such a turn-out is not only a source of health and enjoyment in the beginning of practice, but getting it indicates that your practice is growing. Many persons consider success the chief test of merit, and prefer a much-employed riding physician to the worn pedestrian. This is one of the reasons why any one can RIDE into a full business much quicker than he can walk into one. Besides, the inexperienced public, with nothing else to judge by, infer that a physician who finds a carriage necessary must have an extensive and successful practice, else he would not require and could not afford one.

If you unfortunately have a bony horse and a seedy-looking, pre-Adamite, dust-covered, rust-eaten kind of buggy, do not let them habitually stand in front of your office for hours at a time, or drive a vehicle covered with last week's mud or clay, as if to advertise your poverty, lack of taste, and paucity of practice.

If you have two horses, and two only, it is better to drive singly, that one may be resting while the other is working. Driven thus, two good, well-kept horses can surely carry you to as many patients as you can attend.

If a pair is driven, they should be first-class; for it is better to use one genteel-looking horse to a handsome phaeton, than a shabby pair to a rickety-looking vehicle.

Many physicians have a modest monogram or their initial letter put on their bridle-blinds or carriage-panels. Such designations, when within bounds, are both genteel and ethical.

Either have a person with you to mind your horse, or tie it before entering your patient's house, that you may not be wondering what it is doing, or running to the window or out at the door at every noise, to see whether it has started off with the carriage, as if your mind were more on it than on the patient. When possible, it is better and safer to keep a driver.

While it is perfectly fair and proper to seek reputation by all legitimate means, and to embrace every fair opportunity to make known your attainments, avoid all intriguing and sensational scheming to obtain practice. Attempts to puff yourself, your cases, your operations, or your skill, into celebrity, by driving ostentatious double teams, or having a flashily liveried driver, odd-shaped or odd-colored vehicles, close carriages, conspicuous running-gear, loud monograms, flashy plumes, or oversized initials on harness or carriage-panels, or blazed-faced, peculiar-looking horses or ponies; or pretending to be overrun with business by driving unnecessarily fast, as though the devil were in chase, book in hand, attempting to read as the carriage whirls and jolts along; or having yourself unnecessarily called out of church, at the stillest and most solemn part of the service—

"You assume a hurry, if you have it not;"

and, worse still, affecting odd-style or extra wide brim hats, long hair, and heavy canes; or showing everybody affected kindness or meddlesome attention; and other vulgar, mean, and dishonorable attempts to pass for more than one is worth, to get business,—all generally fail in their object, and are looked upon by many as either an illegitimate, unethical display of artifices and tricks, or the efforts of a small mind or of a weak and ignorant Dr. Sham or Dr. Gullumall to hide a lack of ordinary skill, or to get oneself talked of, and actually sometimes bring him who affects them into ridicule and disrespect.

"Full many a shaft with purpose sent,
Finds mark the archer little meant."

Be cautious not to thus belittle yourself, but strictly avoid ostentation and every peculiarity of manner, dress, office arrangement, etc., calculated to make you offensively conspicuous, and excite ridicule, disrespect, or contempt. On the other hand, however, if you are bashful, shame-faced, diffident, and lacking in aggressiveness or deficient in tact, you will never prosper until these disadvantages are overcome.

In medicine, reputation that comes easily goes easily. Accident or trick may bring one into notice, but they cannot sustain him, and he is finally estimated at his true value. The best reputation is that acquired by a display of talent and merit. If one is tossed into reputation he does not merit he will surely sink again to his true level. Even if you get reputation for distinguished abilities by superior talent, and desire to sustain it, you must still work hard,—

"A great reputation is a great charge,"—

and from time to time present additional ideas and show new proofs of possessing talents and intellectual strength.

It is customary and proper to give simple notice of removals, recovery from prolonged sickness, return from long journeys, etc., in the newspapers, but it is neither legitimate nor creditable to announce your entrance into practice; nor to advertise yourself generally in newspapers, nor to placard barber-shops, hotels, etc. Puffing yourself, your cases, your apparatus, your skill, or your fame, through the medium of the press, and winking at being puffed and applauded in the newspapers, are quackish, stale, unprofessional, dishonorable, and on a par with Dr. Hugh DeBrass and his speckled-horse plan. A proper pursuit of medicine will imbue you with loftier sentiments and engender nobler efforts to gain public attention and to get yourself talked of, and will spur you to build your fame on much stronger foundations.

Cultivate the true art and spirit of professional manner and deportment. Much of your usefulness and comfort will depend on it. But do nothing to gain popular favor that does not accord with both the letter and spirit of the ethical code. Independent of the degradation you would feel, it would not pay to trust, for business, to tricks of any kind; for the veil that covers such attempts is generally too thin long to hide the real motive or to turn aside ridicule.

You will be more esteemed by patients who call at your office, for any purpose, if they find you engaged in your professional duties and studies, than if reading novels, making toy steam-boats, chasing butterflies, or occupied in other non-professional or trivial pursuits; even reading the newspapers, smoking, etc., at times proper for study and business, have an ill effect on public opinion, which is the creator, the source of all reputation, whether good or bad, and should be respected; for a good reputation is a large, a very large, yea, sometimes the chief part of a physician's capital.

It is very natural to expect your near medical neighbors to pay you a visit of courtesy after you commence practice, or change your location; for the purpose of establishing reciprocal and friendly intercourse, whether previously acquainted or not; but if they fail to do so, it should not be too quickly construed as discourtesy or ill-will, for it may be due to their position of doubt concerning your being a regular physician; or they may deem it your duty to make the first call, to announce your intention to practice in the locality, and to tell of your honorable business hopes and ethical intentions, and to ask for kindly, courteous treatment; or, they may wish time to scrutinize your principles, or your character, or your conduct, qualifications, temper, etc. The very best of men are sometimes the slowest to make friendly overtures.

There is a very great difference between the case of an additional physician starting in a community or a neighborhood, and an additional person being added in almost any other business. The demand for other things can be increased, but the demand for physicians is limited; so that a new physician must create his practice by securing this patient, then that, then another, from other physicians. Every family the new competitor adds to his list during his first years of practice must leave or be diverted from that of some other, who may have attended it long enough to almost deem it his private property; and, of course, the loser does not enjoy the loss of his old patients, for there is a little of the old Adam and love of monopoly still left in a man, even though he does practice medicine. The older practitioners are, therefore, naturally very apt to feel a tinge of jealousy, and to be watchful of, if not captious toward, Dr. Newcomer; and when they see him crowding himself in, interlocking and overlapping them, much as we see a new passenger push into an already crowded street-car, they are apt to look upon him as a presuming antagonist and opponent, and, as self-preservation leads every man to prefer himself to his neighbor, unpleasant animosities and feuds are apt to arise, either among those who are well disposed or otherwise. Beware of these differences, and try to nip them in the bud.

There is a proverbial rancor and bitterness of spirit about medical antagonisms and medical hatreds, some of which terminate only with life; avoid them as far as lies in your power, and endeavor to be in amicable and brotherly relations with the physicians of your neighborhood; and should you ever feel that you have cause for complaint against a brother physician let him know of it, and give him an opportunity to explain and defend his action, or to acknowledge his error, if he is in error; then, if you disagree, refer the case to mutual professional friends for adjustment; or, if you have been too badly treated to admit of these, you may feel compelled to drop intercourse and pass him silently. Remember, however, that nothing is more disagreeable than to have enmity and a rupture of intercourse with those we must often face.

It is natural for established physicians to regret the advent of another medical aspirant; and some are suspicious, cold, sensitive, and hypercritical toward every new-comer; because the stranger, in coming, must exert a perturbing effect on the professional business of those already established. His coming makes more workers, and, if he is skillful, actually makes less sickness, because the spur of rivalry, constant and sharp, stimulates each person to try to get all curable cases well, not only surely, but quickly. Sickness, both in amount and duration, is decreased, because skilled laborers have increased. There is, of course, no greater number of cracked skulls, mangled limbs, cut fingers, ague, fits, or medical cases of any kind, than before Dr. Last came. He must, therefore, draw his share of the loaves and fishes from the others.

Read how eager young Absalom was to push old David from his throne, and study the manœuvres of that ungrateful bird, the cuckoo; how the fostered cuckoo hurls all the other birds from their maternal nest after its cunning mother has been unwisely allowed to deposit an egg, and their parent has watched and nourished it until it is strong enough to show its ingratitude by hurling the rightful owners out, and you will realize why Dr. Elder, Dr. Bigbiz, Dr. Nopolizer, Dr. Duwell, Dr. Kurumm, and other old and prosperous physicians dislike to see new Richmonds gain a foothold in their section, and under their very noses effect an entrance into their families. Competitive practice does not necessitate jealousy or enmity; but self-preservation is the first law of nature, implanted by the great Creator of us all; when it is endangered, every human bosom feels the same impulse.

Bear in mind, honest, conscientious, courteous rivalry between physicians is advantageous to the public, because it creates a spirit of emulation and compels each to try to be skillful and successful in practice; and that if your opponents look to their own good, and do all they can for themselves in a fair, equitable, well-directed manner, you have no right to complain.

Your first efforts in practice will bring you into contact and contrast, perhaps also into collision, with the other practitioners of your vicinity, and then you can each learn what the other is.

Be not boastful or intrusive, but if you are conscious of any superior aptitude or intellectual power, or are ahead of your brethren in any essential quality, or eclipse them in talent or experience, let mere matters of display remain secondary, and depend chiefly on your solid merit for success. This is more durable, less expensive, more in harmony with the views of sensible people, and will help you more in climbing toward the top, and, when you get there, will be the surest means to keep you there.

Every one on the face of the globe tries to be wise for himself, and studies his own interests, and desires his own advancement; therefore, do not hesitate to embrace fully every accidental or natural advantage of birth or wealth, or the favoritism of influential patrons or the recommendation of powerful friends, if honest and ethical.

You will find that intellect, genius, temperance, correct personal habits, and other excellent qualities will all fail to make you successful, unless you add ambition, self-reliance, and aggressiveness to them; but in your efforts to advance you should take care not to incur the reputation of being a sharper or of being tricky. If the balance were struck, it would probably be found a great deal harder for a physician to worm and intrigue his way through life, by ingratiating and manœuvring, than to struggle along with honesty and industry. Determine, therefore (under God), that in your efforts you will act like a man, from your diploma to your death-bed; that you will begin well, continue well, and end well; and will do nothing that is criminal, nothing that will not stand the strongest sunlight and the severest scrutiny; nothing for which you would hesitate to sue for your fee; and, if necessary, to stand up before a judge and jury to claim it; nothing, in fact, that you cannot approve of with your hand on your heart and your face turned upward.

## CHAPTER II.

"He who does the best his circumstance allows,
Does well, acts nobly, angels could do no more."

THERE has been of late years a large, annual addition to our already overcrowded profession, and the doctor-making colleges of the United States, with their tempting inducements to students,—small fees, condensed lectures, quizzes, "loading up" at the heel of the session from "compends," "epitomes," "vade mecums," and "multum in parvo" guide-books, and evenings at grinding clubs; with the two short courses of lectures required for *astonishing the professors* in the green-room, by accurately repeating the majority of their own sapient sayings, and thereby obtaining the M.D.,—are now manufacturing annually more than four thousand graduates, besides the medical immigrants representing all nations who reach our shores from abroad, already dubbed M.D., and prepared to enter at once upon practice. The result is that, if it requires a population of 1800 to support each physician, and if every physician must have a paying clientage of 1000 or 1200 persons to enable him to live and thrive, there are now in every American community more than twice as many physicians as are required by the professional work.

Yea, every city, town, hamlet, and village, every cross-roads, every nook and every corner, everywhere in our land, can now boast a physician or two. Canada has but one for every 1193 inhabitants, Austria one for every 2500, Germany one for every 3000, Great Britain one for every 1652, France one for every 1814, Italy one for every 3500, while we of the United States, blessed (?) in physicians as in everything else, have, counting both regulars and irregulars, one for every 600, and druggists in proportion. If there were only a few more than needed to fill vacancies caused by death and increase of population it might be wholesome, and would allow the public a choice, but with such an overproduction as this there is not professional work enough to employ all, and many worthy aspirants must necessarily languish, and those who do flourish must do so by great skill, great tact, or great industry. Another result of issuing diplomas so freely is that diplomas are now far down in public estimation, and are not received as evidence of their owners' competency either by army or naval examining boards, or even by State licensing boards.

The doors to the Æsculapian temple are open,—too open to every variety of individual,—and all kinds are rushing in, and you will be unusually lucky if you encounter none who are maliciously antagonistic. You will not only meet Professor Loveall, Dr. Fair, Dr. Ettykett, Dr. Warmgrasp, and Dr. Dove, but Professor Crank, Dr. Oblique, Dr. Sneerer, Dr. Crusty, Dr. Quackit, Dr. Squabler, Dr. Frigid, and Dr. Spitfire are also about, and may be encountered in unfriendly collision.

Bear this fact in mind, and avoid all manifestations, and, if possible, all feelings of petty jealousy, and let your conduct be affable and frank, fair and square to everybody on all occasions, and strive, in your daily life, to build a reputation for professional probity that will excite the respect of all, whether friend or foe, and convince them that you are incapable of any dishonorable act.

Avoid all quarrels, bickerings, and disputes with your medical brethren, and be ever ready to yield a point, where it involves no principle, rather than engage in controversy and contention; and if ever a question arises between you and a brother physician that you cannot settle yourselves or by the code of ethics, submit it to the decision of mutual friends, but never begin to retaliate or make reprisals, and avoid all innuendoes and sarcastic remarks to the laity about opponents who have offended you. Exhibit a total absence of professional tricks, and resolve, once for all, that you will remain and act as a gentleman, even under provocation, whether others do so or not. Fail not to practice the golden rule, and "do unto others as you would have them do unto you," and trust the balance to time. Medicine is an honorable calling; resolve that it shall be no less so by your adopting it.

Remember, too, that honor and duty require you to do right not only because it is good policy, but because it is right. Do not, however, be so trusting as to "look for wings on a wolf," or expect exact justice from rivals and personal enemies in return; for, were you as chaste as Diana and as pure as the falling snow, you could not escape misrepresentation by evil eyes, wicked hearts, and deceitful tongues.

Like every other physician, you will have your friends to extol you and your enemies to condemn and decry you, and although you can neither stop the latters' tongues nor prevent all unfavorable public criticism, yet you must take care that nothing be permitted to blast your reputation for upright, honorable conduct. Charges against your skill, unless very gross and damaging, had better be left unnoticed, or passed over with indifference; even though it reaches your ears that some Littlewit, or Grundy, or Glibtongue has said he has a total lack of faith in you, and would not call you to attend his ailing cat or dog, such sarcasm need not disturb your equanimity, nor be taken as personal; remember that such remarks are simply individual expressions of lack of faith in you professionally. Such things are said about every physician in the world, and, although they grate harshly when they reach the ear of him to whom they apply, they are quite different from personal libels, or such as bring your morals or integrity into question,—charges of being a swindler, or a drunkard, or an adulterer, or a seducer, or a murderer, or an abortionist, for example.

Never boast of the number of cases you have; of your remedies, operations, and wonderful cures; or of the surprisingly large amounts of your collections. All such things are apt to create envy, jealousy, disbelief, adverse criticism (Professor Pufhimself or Dr. Hornblower), and other hurtful results.

Also avoid talking about yourself, or telling from house to house how terribly busy you are, and of your numerous bad cases, and claiming to save the lives of all who do not die. Indeed, it is better to say but little in regard to your own merits, either in the way of exaggeration or depreciation, and to relate nothing at all to laymen about any case but the one before you; phthoothorn bragging will not enhance your merits with sensible people, and if you really have extra cases and extra skill, or are a great anatomist or eminent surgeon, people will be sure to find it out in other ways. Also keep your business affairs and your money matters to yourself, and avoid the habit of talking to people about your collections, bills, etc., unless it be to a person about his own bill, or you will soon get the reputation of thinking and talking more about money matters than anything else.

As a physician, you will require a good address and varied talents, for you must come in contact with all kinds of people. An intelligent readiness in adapting yourself to all classes sufficiently for the requirements of your profession is an invaluable faculty, and one in which most physicians are sadly deficient.

In addition to professional knowledge, you should make yourself fairly conversant with general scientific subjects that tend to exercise the reason rather than the memory, and also with general and polite literature, that you may acquire ideas, a nice discrimination of words, and improved power and facility of expression, and so put yourself on a conversational level with the cultured classes with whom you are likely to be brought into contact. In fact, among intellectual and educated people, good conversational powers and broad culture often actually produce a higher opinion of a physician's professional ability than is really possessed. Besides,

"Wisdom is the sunlight of the soul,"

and there is a perpetual delight in the possession of knowledge. Therefore, keep your dictionaries and encyclopædias at your elbow; patronize them freely, and, when your reading or musing excites your curiosity on any subject, turn to them and be informed. They are very convenient and useful in looking up facts and opinions when you have but a few moments to devote to an inquiry.

> "We live in thoughts, not breaths.
> He most lives who thinks most."

One who can neither conjugate *amo* nor decline *penna* may reduce a dislocation, adjust a fracture, tie an artery, or prescribe a drug as skillfully as the Latinist can; yet a good (classical) education, and the mental images, ideas and discipline that follow, although not indispensably necessary to the acquirement of skill, experience, and success as a physician, are powerful elements in the professional struggle. Therefore, if you have begun late in life, and are defective in early training, be not cast down; but, to rid yourself of the charge of illiteracy and misapplication of words, make up the deficiency by dint of study and self-education, as fully as possible; otherwise, it will make you ashamed of your want of knowledge, and either keep you hid among the nonentities of the profession or perpetually debar you from obtaining more than a limited elevation in it.

Indeed, without educational and other qualifications you can no more enjoy social or professional rank, or reach true eminence, than a pigeon can fly upward with but one wing. The true secret is to be qualified for advancement; besides, without a fair education you will be continually exposed to ridicule for your ignorance or vulgarism by persons who are, perhaps, very much your inferiors in those peculiar gifts of heaven,—genius and sound common sense. But while a physician cannot know too much, I strongly doubt the wisdom of frittering away, after practice is begun, a disproportionate amount of time on educational frivolities and school-boy subjects, or giving them more time than recreative attention allows. Nor is it wise to give special attention to higher mathematics, the fine arts, the great classics, zoölogy, comparative anatomy, mineralogy, botany, Egyptology, geology, conchology, or other collateral studies, while yet imperfect in the practical and essential principles of medicine, because simultaneous attention to multifarious subjects prevents concentration of thoughts, and naturally divides and distracts one's mind, and prevents one from pursuing the strictly needful studies with his full strength. Do not attempt to grasp more than you can hold, but pursue whatever you do undertake with manly determination and continuity of effort.

The plan of forcing themselves tenaciously to pursue aims of a practical character constitutes the peculiarity of most men who rise much above the ordinary level and succeed in an eminent degree. This is not only true in medicine, but in any calling. I once knew a person who by accident lost his leg at the middle of the thigh; previous to this he was but an ordinary swimmer, but afterward the fact of his having only one leg attracted special attention to his swimming. Seeing himself thus observed stimulated him continually to do his best, which made him more and more expert, until eventually he became the best swimmer I ever saw, because the most ambitious.

A knowledge of Latin to even a limited extent is of inestimable value. If you are not a scholar, and have not had the advantage of embracing it in your early education, you should not fail to employ some Latin scholar to teach you at least as much as you need in your practice; you can get one at small cost by advertising anonymously in any daily paper. He can, with the aid of a Latin grammer (Gildersleve's Latin primer is excellent) and a dictionary, teach you in a short time sufficient of the outlines of the Latin language to enable you to understand the etymological import and pronunciation of words, phrases, and technical terms, and to write prescriptions, etc., correctly, and thereby lift you above a feeling of abashment at your deficiency in this obviously important particular, give a constant sense of security, and afford perpetual satisfaction. No matter where you get your Latin, so you get it somewhere. Ability to write prescriptions in correct Latin, also, naturally assists in creating respect, or, rather, in preventing unfriendly criticism and disrespect, in the minds of your fellow-physicians, the druggists, and others. Besides, all laymen suppose that every physician understands some Latin, and if they find him ignorant of this they naturally think him equally so in everything else.

Many people really believe we write prescriptions in Latin in order to mask their ingredients. The true intent, however, is to give every article (and every quantity) a concise and specific title, and to point it out in such a manner that when we call for it in a prescription we may get *it*, and nothing else, thus making mistakes of meaning between the prescriber and the compounder impossible; besides, the Latin names of drugs are the same in America, Europe, and elsewhere, and can be read by the scholars of all nations, while the common names, sugar of lead, laudanum, black wash, etc., are liable to differ with each nation and locality. Thus, aqua is water in Baltimore, and is the same in Paris, in Calcutta, and in St. Petersburg. Latin is a dead language, belonging to no modern nation, and therefore fixed, and not subject to mutations. It is not only perfectly accurate, but, by long usage, is in high repute.

A rudimentary knowledge of Greek is also useful, as from it have been formed three-fourths of the compound terms employed in the medical and other sciences. Indeed, Latin and Greek have furnished the materials for building up the language of the various sciences for more than two thousand years. The meaning of the terms semi-lunar and dys-uria are as plain and descriptive to those who understand Latin and Greek as the words milk-pail and steam-boat are to those who understand English.

In using the Latin names of medicines, diseases, muscles, etc., be consistent. Adopt either the broad English or the (Roman) Continental pronunciation, but, whichever you adopt, be careful to use it invariably and correctly. You may acquire a correct pronunciation of the various medical terms by frequently consulting a dictionary, of which there is none better than Dunglison's latest edition.

German is another of the world's great languages, and an acquaintance with it is not only pleasurable and a means of intellectual improvement that costs but little money, but it will assist you greatly with the industrious, faithful, and thrifty Germans, among whom you will find many of your most honest and grateful patients. Determine to get at least a smattering of it early in your career. If you speak German, it is well to mention the fact on your cards.

Remember that no one can learn to speak the German or any other language unless conversation enters largely into his teaching; he must learn it through his ears, as well as through his eyes.

You will find that many foreigners prefer an American physician who can speak their language to one who has come here from their own country, and have more confidence in him, because, being a native, they know he has spent his whole lifetime here, and they reason that, although the great principles of medicine may be taught and learned anywhere, he is by experience more familiar with the diseases that exist in our climate, the peculiarities of the vicinity, and the modifying influences of our seasons, diet, and modes of living.

A German, Frenchman, Spaniard, Italian, or Bohemian will often be delighted to find a physician in an English-speaking community with whom he can converse in his own tongue. Foreigners often pay much more liberally than natives, and usually treat the physician with much greater respect.

A physician is at perfect liberty to state on his cards and signs that he speaks French, Italian, Spanish, Bohemian, German, or any other foreign language; and such a statement should, if made, be in the language of the people for whom it is intended.

Accustom yourself to use current and correct orthography, and to write, not with a scrawling hand, in a zigzag or the worm-fence style, but in a good, neat, distinct, school-day hand. Write every prescription as though critics were to judge you and your penmanship by it; each ingredient on a separate line, the principal article, or the strongest drug on the first, adjunct on the next, and vehicle on the last, unless you have some special reason for inverting them. Such methodical system insures well-balanced prescriptions, and engenders the respect and favorable criticism of those into whose hands they chance to fall. Also, take care to conform your prescriptions to the changes that are from time to time made in the names of the officinal articles of the pharmacopœia by authorized bodies and nomenclators.

Strictly avoid prescribing incompatibles, both chemical and physiological, such as the combination of chlorate of potassium with tannic acid or with sulphur, nitrate of silver with creasote, etc., which are explosives, and may blow up either the dispenser or the patient. Charcoal is a simple thing, sulphur is another, and saltpetre is still another, but put them together and you have gunpowder, which is not simple, and, unless that potent agent is intended, look out. Although the list of incompatibles is a long one, you will do well to learn it thoroughly, otherwise you will subject yourself to the sarcastic remarks of the pharmacist, and possibly to whispering doubts and disparaging innuendoes. Remember, however, that some medicines, though physiologically incompatible, are not therapeutically so, as under certain circumstances you may combine them so that they may favorably modify each other, as morphia and belladonna, acetate of lead and sulphate of zinc, etc.

Instead of writing prescriptions three inches in length, it is better to use a single remedy, or, if two are indicated, to alternate them, unless you know they are compatible and will not make an unsightly mixture.

Again, your prescription is always the expression of your opinion and of your skill in a case:—

"The mind is the man."

Therefore, try to make every one you write show on its face that you have prescribed with a definite purpose, to meet some clear indication.

Be careful that abbreviations of names, manner of writing quantities, etc., leave no room for mistake or inexactness in dispensing, and make it a rule to read carefully every prescription after you finish writing it.

It is scarcely necessary to add that, while the distinctive names of the several ingredients in a prescription should be written in Latin, the directions for use, *i.e.*, all that follows the S. (signa), should be in English, as they are intended for the guidance of the patient.

Remember that the cloven-foot ℞, that is placed at the head of every prescription (præ, beforehand; scriba, to write), although originally the astrological sign for Jupiter (♃), and for ages placed by the ancients at the head of prescriptions, to invoke the aid of the God of Thunder, is now used merely as a symbol to represent the Latin word Recipe (take thou).

While it is proper, strictly speaking, to commence every word, after the first, in the names of the articles in your prescription with a small letter, *i.e.*, Liquor potassii arsenitis, yet many physicians purposely begin each with a capital, chiefly because it looks well, and also renders the word less mistakable.

Sign either your name or initials to every prescription you write, that the pharmacist may recognize its writer; to such as are likely to be compounded by pharmacists who know you well the initials will be sufficient, but, to all that are likely to be put up by those who know you not, put your full name.

In prescribing, it is a bad and injudicious habit to adopt a routine practice, or slavishly to follow your own, or anybody else's, stereotyped formulæ for certain diseases. You should invariably adapt your remedies to the case, instead of heedlessly picking out a ready-made formula from your collection as you would a hat in a hat-store. One formula, for instance, for the several forms of diarrhœa, is about as apt to suit every case of relaxed bowels as one coat is to fit every man in a regiment.

Remember that medicine is a mass of facts, and that he who best interprets these facts is the best physician, and that skill in practice consists not only in diagnosis, prognosis, and prescribing medicine, and in knowing what one can and what one cannot do, but is the combined result of all the powers that the physician legitimately brings into the management of cases. In other words, the skillful use of drugs is but *one* of many elements that make the unit of medical skill. You must also study mankind as well as medicine, and remember, when working on diseased bodies, that they are inhabited by minds that have variable emotions, strong passions, and vivid imaginations, which sway them powerfully, both in health and in disease. To be successful you should fathom each patient's mind, discover its peculiarities, and conduct your efforts in harmony with its conditions. Let hope, expectation, faith, contentment, fear, resolution, will, and other psychological states be your constant aids, for they may each at times exercise legitimate power, and impart the greatest amount of good to the sick. It is not length of time in practice, but observation and reflection, that teach one to measure human passions and emotions; and if you are not a keen observer of men and things, if you cannot read the book of human nature correctly, and unite knowledge of physic with an understanding of the effects of love, fear, grief, anger, malice, envy, lust, and other hidden but strong passions that govern our race, you will be sadly deficient even after twenty years' experience:—

"Hair gray, and no brains yet."

Professional fame is a physician's chief capital; ambition to increase it by all legitimate means is not only fair, but commendable. After you attain this, you will not be apt to lose either it or the practice it insures, so long as you are sober, decent, and discreet in conduct, and have the physical health to endure the watching, fatigue, and exposure incident to our business.

There are two kinds of legitimate reputation a physician may acquire,—a popular or common one with the people, and a higher professional one with his brethren. These are often based on entirely different grounds, and are usually no measure of each other; a few of the most excellent, with loftier ambition, struggle earnestly for the latter, while the mass are striving for the former, chiefly because, being altogether practical, it requires less skill, talent, and study to acquire, and, also, because it is more profitable. Many such avoid all great scientific labors and controversies, and, having little or no public life, remain shut up within themselves, moving about quietly and almost unobserved except by those whom they attend; consequently, a knowledge of their habits and doings is confined to the domestic bedside and the narrow circle of their private practice, and the degree of their skill and experience always remains somewhat unknown and mysterious.

Without one or the other variety of reputation no physician can reap the honors or rewards which are the objects of his ambition, whether that be the acquisition of money, the desire of usefulness, or the love of fame. You should strive to acquire both varieties.

One fact that you will notice is, that the public naturally prefer a full-of-health, ever-ready physician to a delicate or sickly one, and ailing physicians often conceal the fact that they are sickly or that their health is failing as much and as long as possible, well knowing that the competition in our profession is now so great that for every person whose powers fail ten are ready, with fresh strength, to take his place, and that, if reports of their ailments become current talk, the public will believe that solicitude for their own condition will absorb it from their patients, and they will be abandoned as unreliable and unfit to practice, and their business will be thereby injured or ruined.

After you have practiced awhile and discovered what your chief deficiencies are, and determine exactly what course you ought to pursue, if you will spend a few months in additional study of the great principles of our science in some of the great American or European hospitals, and then return and settle down, it will be of tenfold benefit to you in more ways than one.

A discreet tongue is a great gift and a great aid to success. When elopements, seductions, rapes, confinements, or abortions; or the scandal about Dr. Bigscamp, or Rev. Mr. Blacksheep, or Miss Oilyeve, or the ignobleness of the pedigree of Mrs. Butterfly, or the secret history of Miss Pride, or the wrecked and wretched greatness of Mr. Pomp, or the adulteries or intrigues of Mrs. Freelove, or the evil reports about this virgin, that wife, or the other widow, are being talked of, perhaps in terms that decency would require to be printed only with initial and terminal letters, with a dash between, you should have a silent, or at least a prudent, tongue; all you say on such subjects will surely be magnified and repeated from mouth to mouth, and its results will be a permanent injury to you. The position of the gossiping physician has ever been a very bad one, and he is sometimes called to unpleasant account.

Take especial care, while in contact with tale-bearers and scandal-mongers, or scandal-loving crowds, to keep the conversation on general or abstract and legitimate subjects, and determinedly avoid descanting upon individuals and private affairs, or what somebody, or a coterie or clique of somebodies, has said.

Be careful, also, to note the great and never-failing advantage that refined people, with virtuous minds, pure thoughts, and courteous language, have, in every station of life, over the coarse and the vulgar; and in view thereof let your manner, conversation, jokes, etc., be always chaste and pure. Never forget yourself in this particular, for nothing is more hurtful to a physician than the exhibition of an impure mind. School yourself to avoid all and every impropriety of language and manner, and never allow yourself to become insensible to the demands of modesty and virtue. Chasten every thought, weigh well every word, and measure every phase of your deportment,—especially that which concerns the fair fame of woman,—and let your treatment of all females be refined and delicate, if you would succeed fully, especially if gynæcology and obstetrics be the one great aim of your ambition. A lewd-minded physician who indulges in double *entendres*, coarse ambiguities, vulgar jokes, jocular innuendoes, and indelicate anecdotes about the sexes—

"To reflect on women ever ready"—

with other men or with coarse women, even though he poses as a gentleman, is sure to be shunned, and the reason therefor made the subject of gossip and passed from one to another in social whispers, till it reaches the purest and best of the community. Thoughtful people of both sexes everywhere rightfully regard such libertines as being far more amenable to criticism, and far more dangerous to admit into the bosoms of their families, than rough-mannered believers in social purity who gamble, drink, or swear.

"Immodest words admit of no defense,
For want of decency is want of sense."

Study the art of questioning, and when it devolves on you, in the course of professional duty, to ask questions on delicate topics, or to broach very private subjects, do so with a chaste, grave simplicity,—neither too direct on the one hand, nor with too much circumlocution on the other.

Physicians are made in the colleges, but tried in the world. Your personality and deportment in the presence of patients will have much to do with your success. Blessed is the physician who has the gift of making friends. A pompous, or cold, or cheerless, heartless or indifferent manner toward people; or a studied or sanctimonious isolation of one's self from them socially; or failure to recognize would-be friends on the streets and elsewhere, as if from a lofty independence, or as if they were inferior mortals and beneath you,—

"I am resolved on death or dignity,"—

often gives unmeant offense, and tends to destroy all warmth toward a physician, and usually causes their hapless possessor to fail to inspire either friendly partiality or faith; and a physician who cannot in some way make friends or awaken faith in himself cannot fail to fail. The reputation of being a "very nice man" makes friends of everybody, and is, with many, even more potent than skill. To be both affable in manner and skillful in action makes a very strong combination,—one that is apt to waft its possessor up to the top wave of professional success and repute. If, moreover, he be especially refined in manner and moderately well versed in medicine, his politeness will make him a troop of friends, and will be professionally more effective with them than the most profound acquaintance with histology, microscopic pathology, and other scientific acquirements.

If your manners and conversation are of the gentle, soft, and tender kind, that win and conciliate rather than repel children, it will be fortunate, and probably will put many a dollar into your pocket that might have gone to some irregular. Such habits as fondling and kissing people's teetsy-weetsy children, or carrying them pockets of candy, however, are liable to be misconstrued into an effort to secure the good will of the parents for selfish motives, and should therefore be avoided.

Cultivate a cheerful mental temperament; gentle cheerfulness is a never-failing source of influence. It is a magnetic nerve tonic and stimulant; it diffuses sunshine, cheers the timorous, dispels the deadening fogs of hopelessness, encourages the despondent to look on the bright side, and comforts the despairing. The science of medicine, contrary to the general belief, is not a melancholy, sombre, mournful profession, but a bright, cheerful one. The sincerely grateful faces you will see and the "Thanks to God!" you will hear while completely curing some poor fellow-creatures and relieving others of pain and ailments, and allaying fear and administering comfort to the minds of multitudes of others, will make you realize your usefulness and the great good your noble, humane, and beneficent profession enables you to confer on suffering humanity,—the contemplation of which should make you cheerful and happy, and satisfied with yourself and your elected life-work, in spite of the many contradictions and disappointments you are subject to in practice.

Bear in mind that the physician's visit, being the chief event of a sick person's day, is eagerly watched for, and let no ordinary engagements interfere with your punctuality in making it; also study to acquire an agreeable, courteous, gentlemanly, and professional manner of approaching the sick and taking leave of them. There is an art, a perfection, in entering the chamber of sickness with a dignified yet gentle manner, that clearly evinces interest and a determination to master the case,—in asking the necessary questions, in making the requisite examination, then carefully and wisely ordering the proper remedies, and departing with a cheerful, self-satisfied demeanor that puts the patient at his ease, and inspires confidence on the part of himself and his friends, and a belief that you can and will do for him all that the science of medicine enables you to do. The personal appearance, the walk, the movements, the gestures, the polite bow, the well-modulated voice, the language, the natural mode of intercourse, and the elegant and instructive conversation of some physicians are as cheering and confidence-inspiring, to the sensitive nerves of the sick, as a sunbeam on a May day; the manners of others, as rude, coarse, cold, heartless, indifferent, and repulsive as a March wind.

Familiarity with the many little details of the sick-room—including the necessary art of applying bandages, making beef-teas, gruels, mustard plasters, poultices, etc., and with dressing wounds, passing catheters, reducing herniæ; getting a fish-bone from the throat, a splinter or a needle from the hand, or a mote from the eye, or teaching the nurse how to prepare the obstetric bed; seeing that those working subordinate to you do their duty, and various other minor duties that you may be there incidentally called on to perform or direct—often do more to create a favorable impression than your pills and powders. Indeed, it is to a very great extent by minor matters that watchful nurses and other habitues of the sick-room will judge you.

As a physician you should be hopeful, and not indiscreetly abandon cases usually considered hopeless. Hope creates ideas, generates new expedients, brings up useful reflection, and leads to fresh endeavors. Indeed, it has been said that the only way to get cured, and render impossibility possible, after a physician loses hope and gives you up, is to give him up.

The faculty of keeping hope and confidence alive in the bosom of the patient and of his friends is a great one, and the look with which you meet them has much to do with this; a bright, fresh, thoughtful countenance, and an easy, cheerful, soothing, professional air and manner are powers that will well-nigh always impart tranquillity and repose to your patient's mind and carry him with you toward recovery. A cheering word sometimes rekindles the lamp of hope, and does the timorous and despondent as much, or more, good than a prescription. It is, therefore, your duty to gain and retain the confidence of your patient and his friends by all honorable means,— to be gay, pleasant, amusing, serious, or sympathizing, as occasion requires.

It is often very pleasing to the sick to be allowed to tell, in their own way, whatever they deem important for you to know; allow to all a fair, courteous hearing, and, even though Mr. Humdrum's and Mrs. Lengthy's long statements are tedious, do not abruptly cut them short, but endure and listen with calm, respectful attention. A patient may deem a symptom very important that you know to be otherwise, yet he will not be satisfied with your views unless you show sufficient interest in all the symptoms at least to hear them described. When, for want of time, you cannot listen further, or where the recital grows too tedious and becomes too irrelevant, do not lose temper or manifest any annoyance, or check him by a rude order to "stop," but quietly ask him a diverting question about his sickness, or to show his tongue, or feel his pulse, as if completing your examination. Such expedients often serve the purpose with hypochondriacal men, garrulous women, and tedious chronics in general.

To be quick to see and understand your duty, and equally prompt and self-reliant in doing it, as if possessed of inborn acuteness of perception and of intuitive skill, is one of the strongest points you can possess, and gives easy advantage over Dr. Lazi, Dr. Dragg, and Dr. Dullhead, who mildly and formally perform their part, and are as painfully slow, undetermined, and cautious, as if every pebble were a rock and every molehill a mountain. People invariably admire and appreciate the man who can take the responsibility in critical moments; indeed, a bold, prompt act, done at the opportune moment, with steadiness of mind and nerve, if successful, often creates a species of faith bordering on professional idolatry.

Capital operations in surgery illustrate this: the manual parts—expertness with the knife, etc.—are deeply impressive, and receive vastly more praise from the crowd than knowing when to operate and how to conduct the after-treatment. Indeed, the public imagine that the comparative scarcity of surgeons is because but few of our number dare to do great operations. The truth is, almost every physician does minor surgery,—adjusts fractures, reduces dislocations, etc.,—and would prepare to perform capital operations but for the reason that only a few are required to do all there is to be done, and only a very few can live by it. A large city with its hundreds of physicians will have less than a dozen who are prepared to do capital operations, and the majority of these have a great deal more medical than surgical practice.

If you know a patient's ailments so well as to sit down and tell him and his friends exactly how he feels better than he can tell you, he will be apt to believe all you afterward say and do. Mind-reading, or the study of character, is part of your duty. To be many-sided; to possess flexibility of temper and suavity of manner, self-command, quick discernment, address, ready knowledge of human nature, and the happy genius of honestly adapting yourself to varying circumstances and to all people, at the couch of splendor and the squalid cot, are great necessities in our checkered profession. You will meet patients of various and even of directly opposite temperaments and qualities: the refined lady and the hod-carrier, the clergyman and the beer-seller, the aged and the young, the hopeful and the despondent, the bold and the diffident, the profound and the superficial. Let each and all find in you his ideal. Seek to penetrate the character of each, and to become an adept in adapting your manner and language to whoever and whatever is before you.

If you also have the self-command to control your emotions, temper, and passions, and to maintain a cool, philosophic equipoise and inflexible serenity of countenance, under the thousand irritative provocations given to you by foolish patients and their querulous and rude, or fidgety friends, who rile at your coming too early or too late, too often or not often enough, or accuse you of giving the wrong medicine or in the wrong doses, of being too fast or too slow, it will give you great advantage at the critical moment over nervous, quick-tempered, and excitable physicians who unguardedly blurt out with "———,:;——— !!??———*?!!!———!!———?", and will generally redound both to your advantage and credit.

A brusque, tornado-like manner, or eccentric rudeness, is fatal to a physician's success unless sustained by unquestionable skill or reputation. A simple, humane, gentle, and dignified manner and low tone of voice are suitable to the largest part of the community.

"Manners gentle, discourse pure."

Remember that a rough, unfeeling, abrupt, indelicate, sour, or arbitrary manner, as if the heart were a butcher's, or made of marble, is quite different from the serene composure and intelligent sympathy acquired by constant attendance upon the sick and suffering. The former is brutal and unprofessional; the latter is essential to enable you to weigh correctly and manage diseases skillfully.

If you chance to inherit any slight but pleasant peculiarity of character or singularity of manner it will be noticed, and, if not disagreeable, will do you no harm;* but never assume one for the sake of making an impression on the public, for the counterfeit is easily detected by all sensible men and women. Be not only a gentleman, but also a gentle man, and act out your own natural character everywhere and at all times, among the rich and the poor (no man has two natural manners). Besides making himself ridiculous, a physician who assumes a fictitious, mysterious, or rude manner must either be wrong-hearted or weak-headed.

If, moreover, you possess fluency of language, or the gift of conversational power, or gentleness or tenderness of manner, or great natural courtesy, or a never-failing stock of politeness, facility of expression, or a talent for illustrating your points by apt comparisons, or a bold, resolute way of encountering professional puzzles, or of deftly cutting the many Gordian knots so often encountered, it will help you decidedly. If, on the contrary, there is any point in which you are deficient, study and practice until you attain it.

When you reach a patient's house ascertain, if possible, from whoever meets you, his condition, etc., that you may know with what manner to approach him, especially in cases of severe illness, in which it is important to show him no surprise, nor to disturb him with questions that can be avoided.

Never leave a bedside before qualifying yourself to communicate your ideas and opinions of a case to the inquiring friends of the patient clearly, in well-chosen and faith-inspiring language, in case they should be asked.

Never utter a diagnosis or a prognosis in a hurry or flurry. Give your opinion only after sufficient thought, and, if possible, do not afterward change it. Also, to prevent being misunderstood, avoid making varying statements about a case to different inquirers from time to time, but, as nearly as possible, use the same tactful words and apply exactly the

---

*It is said that the *thee* and *thou* of Dr. Fothergill, of London, was worth £2000 per year to him.

same terms to the disease, and even more particularly in consultation cases.

Act toward timid children and nervous patients so as to remove all dread of your visits. Avoid a set, sad countenance, and a formal or funereal solemnity of manner, as these would excite thoughts of crape, hearse, undertaker, and tombstone, and a fear of you, especially if you associate them with a corresponding style of dress. If you have a lengthened, severe visage, simulating

"A walking prayer-meeting,"

or your air and movements are awkward, sombre, severe, smileless or singular, offset them by enforced cheerfulness, suitable dress, etc.

When you visit a patient, neither tarry long enough to become a bore and give rise to the wish that you would go, nor make your visit so brief or abrupt as to leave the patient with the impression that you have not given his case the necessary attention.

To evince an earnest, anxious, tender interest in the welfare of patients, and serious attention to the nature of their disease, and sympathy with their sufferings, as if you were present in mind as well as in body, is another very strong, faith-inspiring quality. To find occasion to assure a sufferer that you will take the same care of him as though he were your "own brother," or, in case it be a female, as if she were your "own sister," or to assure a female in labor that you will be as gentle in making the necessary examinations as if she were an infant, and similar truthfully-meant expressions of sincere sympathy and interest, and letting your conduct be such that they may feel it is so, inspire great confidence, and are often quoted long after the physician has used them.

"A little thing often helps."

The world is full of objects of pity, and it may be that no really busy physician can devote full time and exert his utmost skill in every case that appeals to him, or throw into it his whole heart, undivided force, thoughts, feelings, and intellectual strength; or even feel deep interest in the agonies, the woes, the bruises, the afflictions, and sufferings of every patient to whom he is called; if he did, the endless chain of misery with which he is brought in contact would prove to be too great a strain on his sensibilities, and, through overcare and grief, would soon unfit him for active practice. But you can, and should at least, make a careful examination, in a grave and thoughtful manner, manifest humane anxiety and intelligent interest, and show uniform kindness in all cases, and avoid exhibiting a rough, abrupt manner, unfeeling, thoughtless haste, or chilly indifference in any. Be careful to approach the sick, rich and poor alike, with noiseless step, with kindly, hopeful greeting, and gentle, thoughtful speech. The possession of a feeling of true humanity, or the lack of it, in a physician, can in no way be so accurately judged as in his questioning and examination of the sick; the soothing voice, the tender touch, and the sympathetic feeling tend not a little to soften the pillow of sorrow and affliction.

In examining the sick, be especially careful to use *the professional touch*, and avoid inflicting pain in delicate and painful parts, and assuage their fears and oversensitiveness by assurances that you will not cause any more suffering than is unavoidable, and then proceed to make good your words. He who possesses such manner and tact naturally will not, cannot, fail to gain devoted patients, who will willingly trust and retain him in preference to all others, even though they know his general reputation for skill to be far below that of professional neighbors.

Human life is precious above all on earth; but some persons think that being so often in contact with sickness and death naturally makes physicians less alive to life's value and more callous to suffering than other men, and nothing is more gratifying to all, and especially to such as are interested in one who is lying sick, than to hear the physician expressing a lofty estimate of the value of human life in general, and why the life then at stake is specially valuable, and worthy of an earnest determination on the part of all to save it if possible.

For ultimate success you must, of course, depend chiefly on your skill in curing the sick. You will find, nevertheless, that but few patients—probably not one in twenty—can estimate the amount of technical and scientific knowledge you possess. The majority are governed by the care and devotion you exhibit, and form their opinion and rate your services by the little details of routine attention, which is additional evidence that mere skill is not all that is necessary to make a successful physician.

While civil and urbane to all, without distinction, be especially courteous to female attendants on the sick; for woman, noble woman! as true to duty as Diana, with voice soft, gentle, and low, and the look of heaven in her face, is, and ever will be, the angel of the sick-room,—

"Sweet is her voice in the season of sorrow,"—

and you, as a physician, cannot fail to witness many touching evidences of her tender ministrations; and heroic, unselfish devotion as mother, wife, sister, daughter, nurse, or friend to the sick and suffering, watching around the bedside by day and by night, and ministering with an angel's spirit, even at the risk of her own life.

"Woman, fairest of creation, God's last and best gift to man."

After a patient convalesces, or when it is not necessary to visit him daily, if, when you chance to be attending in his neighborhood, you send to inquire how he is getting along, it will not only give you the desired information, but will also impress him and his with a grateful sense of your interest in the case. Having a sick child taken up for examination, carrying your patient to the light that you may see him fully and examine him carefully, also having his urine, or his sputa, or the blood spat, etc., saved for examination, will not only give you much necessary information as to the patient's condition, but also satisfy him and the family of your interest and solicitude, and of your anxiety to fulfill your duty. A like effect is also produced by paying one your first visit in the morning, or the last at night, or staying, in urgent cases, to see that the medicine produces the desired effect, and such things help to make the cure.

You will find that, in times of sudden sickness and alarm in families, there is a peculiar susceptibility to strong impressions, and kindness and extra attention shown them in such emergencies is doubly appreciated. Often even a single kind expression, opportunely uttered, is long remembered. Indifference, coldness, a slight offense, an inopportune remark, an unlucky word, or an impatient ejaculation, may, on the contrary, sever attachments and terminate friendships that have existed between the physician and the family for years, in as many moments. Many a young physician gains a hold on the hearts of a good family, becomes beloved, and secures the family permanently by the exhibition of good, hopeful intentions, and simple kindness and assiduous attention in those dreadful accidents and emergencies that alarm friends and distress families; and, also, in cases of colic, convulsions, and the like; or by sleepless anxiety and faithful, devoted, and unwearied attention, trying to steer here to avoid this rock, and there to escape that eddy, in cases of typhoid fever, scarlet fever, etc., where, perhaps, life hangs, day after day, as if by a single thread.

A powerful lever to assist in establishing your professional reputation will be found in curing the long-standing cases so often seen among the poverty-stricken. Many of these poor, disease-ridden sons and daughters of poverty are curable, but require greater attention in regard to the details, and a great deal more care, strength, and personal superintendence than old-established physicians, whose time is monopolized by acute cases, can possibly devote to them. If you are seriously in earnest, use your best judgment, and persevere with them until a cure is effected; your special interest and anxious attention will be observed and appreciated; you will be credited with all the prosperous accidents of the case, get the credit of the cure, and gain a host of warm admirers, who will magnify and herald you far and wide as being doubly skillful in making the blind see, the deaf hear, the lame walk, the broken whole again, the senseless well, the weak and debilitated strong, rotten lungs sound again ; and, even though you receive little or no pecuniary reward from them, it will serve as a mental gymnasium, help to train and develop your professional character, show your skill and ingenuity, augment your fame, and educate both your hand and your eye, and school you in the art of recognizing, studying, and treating the very diseases you will daily be called upon to attend all the days of your life; besides, teaching you to overcome the thousand and one embarrassments encountered by the beginner, and bring you eventual success in life. And when success does come, forget not those by whom it came, and with grateful heart be true to all the friends of your struggling years.

> "Thine own friend, and thy
> Father's friend, forsake not."

Take care to promise old chronic cases—that more experienced physicians have pronounced incurable, and annoying and troublesome, but penniless patients, taken for older physicians who wish to discard them—nothing but that you will do your best for them. Never stake your reputation on their cure, and allow yourself plenty of time in speaking of the period necessary for the trial, instead of promising too much, or good results too soon.

You will find it comparatively easy to get practice in the slums and among the moneyless poor, and relatively hard to do so among the wealthier classes. Your practice will probably begin in cellars and garrets, lanes and back streets, among the poorest of the poor, the degraded and the vicious,—even in hovels of filth and vermin, in putrid alleys and fetid courts, where

> "I have counted two-and-seventy stenches,
> All well defined."

You will also be called to attend people who wash with invisible soap, in imperceptible water, and use immaterial towels, who will furnish astonishing illustrations of

"The survival of the filthiest;"

and will also enter dens of iniquity and vice, where you must pick your way through mud and mire amid

"Poverty, hunger, and dirt,"

where your reputation will extend much more rapidly than in comfortable quarters; but, no matter whether in mansion, cottage, or hovel, every man, woman, or child you attend, white and black, rich and poor, will aid in enriching your experience and in shaping public opinion by giving you either a good or a bad name.

"Over rough roads, indeed,
Lies the way to medical glory."

Bear in mind that the wheel of fortune sometimes makes the poor rich, and a few of the more grateful kind then remember the physician who remembered them. Attending the servants of the rich, however, who are sick at their service places, or paid for by the latter, will not improve your reputation much with the powers above stairs: at any rate, not nearly so much as attending the same patients at their own homes, or on their own account. Proud and haughty people who, in their minds, couple you professionally with their servants, garrets, and kitchens are apt to form a low opinion of your status, and of the nature and class of your practice. It is also true that if you attend a poor person gratuitously you will seldom, if ever, be called to his rich relatives; and if Dame Fortune ever makes that poor patient rich, even he may become supercilious, and drop you.

Nor will you find it very satisfactory to attend people who "just call you in to see a sick member of their family," *because* you are attending across the street or in the neighborhood. Those who select you or send for you because they prefer you to all others will be your best and most devoted patients.

The adoption of a specialty, to the exclusion of other

varieties of practice, is successful with but a few of those who attempt it. It should never be undertaken without first studying the whole profession and attaining a few years' experience as a general practitioner.

You are not obliged to assume charge of any case, or to engage to attend a woman in confinement, or to involve yourself in any way against your wish; but, after doing so, you are morally, if not legally, bound to attend, and to attend properly, even though it may be a charity or "never pay" patient. At the same time you have a right, should necessity arise, to withdraw from any case by giving proper notice.

Bear in mind that ethical duties and legal restraints are as binding in pauper and charity cases as in any other, for both ethics and law rest upon abstract principles, and govern all cases alike.

You will probably find hospital and dispensary patients, soldiers, sailors, and the poor, much easier to attend than the higher classes; their ailments are more simple, definite, and uncomplicated, the treatment more clearly indicated, and the response of their system is generally more prompt, and one can usually predict the duration, issue, etc., of their cases with great accuracy. With the wealthy and pampered, on the other hand, there is often such a concatenation of unrelated or chronic symptoms and strange sympathies, or they are described in such indefinite or exaggerated phrases, that it is difficult to judge which one symptom is most important to-day or which will be to-morrow.

With hospital patients, sailors, soldiers, paupers, etc., on the contrary, there are but two classes,—the really sick, suffering from affections of a well-marked type, and malingerers. Such practice is apt to lead the unguarded youth to a rough-and-ready habit of treating every patient as very ill, or else as having little or nothing the matter with him; later, he finds that these crude or possibly overactive methods may answer in public institutions, but they will not suit the squeamish people with nerves tuned to a high key, so often seen in private practice, with indefinite or frivolous ailments, for which the physician trained in a hospital could hardly fail to feel and manifest contempt. Hospital practice is so different from private that but few members of our profession shine conspicuously as practitioners in both spheres. An illustration of this fact is afforded in colleges and medical societies; for the greatest Ciceronian orators in the colleges, and the most fluent debaters and paper philosophers in medical societies, are not necessarily the best or most successful practitioners. The fields are decidedly different, and may lead the mind in different directions. In a word, the possession of didactical knowledge, and the power of applying it at the bedside, are very different things.

Observe and strictly practice every acknowledged rule of professional etiquette. For this purpose it is your duty to familiarize yourself at the very threshold of your professional career with the "Code of Ethics of the American Medical Association," and never to violate either its letter or its spirit, but always scrupulously to observe both toward all *regular* graduates practicing as *regular* physicians. But remember that you are neither required nor allowed to extend its favoring provisions to any one practicing *contrary to* the liberal tenets that govern all regular physicians, no matter who or what he may be.

I am not sure that the medical profession of any other country besides ours has a code of written ethics. Possibly old countries from long custom can dispense with them. But in our Young Land of Freedom the very nature of society requires that physicians shall have some general system of written ethics to define their duties, and, in cases of doubt, regulate their conduct toward each other and the public in their intercourse and competition. Every individual in the profession is, of course, supposed to be a gentleman, actuated by a lofty professional spirit, striving to do right and to avoid wrong, and, even were there no written rules at all, the vast majority would naturally conform to the rules of justice and honor, as far as they understood them. As a consequence, each one's action, when scanned by watchful and knowing eyes, might probably be considered fair in nine doubtful cases out of ten, while in the tenth one might honestly err greatly, or conclude differently from his neighbor on some mooted point, or might be found differing in opinion only from some jealous or captious rival, or crafty, unprincipled competitor, with whom an honorable agreement would be impossible.

The absence of rules for our government would also leave Dr. Allforself and others at liberty to frame their own codes, which might violate all logic and all propriety,—

"The wrong-doer never lacks a pretext,"—

and no matter how equivocal their position, or how crooked and insincere their ways, no one would be in position to prove that they acted from unworthy motives, and not from error of judgment, even in the most flagrant violation of the cardinal, the glorious old-fashioned Golden Rule, the climax of all ethics, laid down by Confucius, and quoted by "Our Saviour," "Do unto another what ye would he should do unto you, and do not unto another what you would not should be done unto you,"—truly, a world of ethics in a nutshell, an ocean of morals in a drop.

The non-existence of a code would also make it possible for Dr. G. to pounce on the patients of Drs. A., B., C., D., E., and F. like a wolf on sheep, and to carry on a regular system of infringements, self-advertising, certificate-giving, and wrong-doing in general, regardless of their rights, and still claim to be as honorable as Socrates, while those aggrieved would have no visible standard of appeal by which the contrary could be proved.

In view of these and many other facts, it has been found necessary to have a code of written ethics for regulating the conduct of physicians toward each other and toward the public generally.

Dr. Thomas Percival, an English physician, in a small book published in London in 1807, proposed an admirable code of ethics, which, excepting a few alterations made necessary by the lapse of time and the advance of medical science, is the identical code adopted by the American Medical Association in 1847, and which from then until now has instructed and governed nine-tenths of our profession throughout this broad land, protecting the good and restraining the bad, just as the Ten Commandments of Holy Writ instruct and restrict mankind in general.

You and every other true physician among us unquestionably owe to it his sacred allegiance.

You and all other physicians are supposed to have studied this code, and to be familiar with its requirements. The moral claim which it has upon you rests not upon any obligation of personal friendship toward your professional brethren, but upon the fact that it provides for every relation, contingency, and occasion, and is founded on the broad basis of justice and equal rights to every member of the profession, shining like the pole-star to guide and direct all who wish to pursue an honorable course; and, being founded on the highest moral principles, its precepts can never become useless till regenerate and infallible human nature makes both codes and commandments unnecessary. It is the great oracle of right and reason, to which you can resort and study the moral aspect of all the subjects that are likely to confront you from time to time, and no better code of moral principles can be found anywhere.

To this lofty code, in a great measure, is due the binding together and elevation, far above ordinary avocations, of the medical profession of our land, and the esteem and honorable standing which it everywhere enjoys.

By its dignity and justness it remains as fresh and useful to-day as when the profession adopted it, more than forty years ago, and if you faithfully observe its canons you can truthfully exclaim: "I feel within me a peace above all earthly dignities, a clear and quiet conscience."

Professional morals are an important part of medical education, and it is as much the duty of every medical college in

America to acquaint its students with the precepts of the code of ethics of the American Medical Association, and to furnish to each of its alumni a copy of it with his diploma, as it is for a mother to familiarize her children with the Ten Commandments.

In our land the code is regarded as the balance-wheel that regulates all professional conduct, and neither Professor Bigbee nor Dr. Littlefish can openly ignore it without overthrowing that which is vital to his standing among medical men. If in the struggle and competition for practice you desire to act unfairly toward your brethren, the code will compel you to do the evil biddings of your heart by stealth; and even then your unfairness will seldom go undetected or unpunished, for the great God of Heaven has declared that "Whatsoever a man soweth, that shall he also reap." Any one upon whom you encroach in an unprofessional manner will feel himself justified in retaliating with your own weapons, and you will reap a crop similar to the seed sown. Whenever you sow a thistle or a thorn you will reap thistles or thorns, whenever a wind is sown a whirlwind will be reaped; whilst the sweeter seeds sown by others will be yielding to them sweeter fruits.

When called to attend a case previously under the care of another physician, especially if the patient and friends are dissatisfied with the treatment, or if the case is likely to prove fatal, be carefully just. Do not disparage the previous attendant by expressing a wish that you had been called in sooner, or criticise his conduct or his remedies; it is mean and cowardly to do either. In all such cases do not fail to reply, to the questions of the patient or his inquiring friends, that your duty is *with the present and future, not with the past.* Inform yourself as to what line of treatment has been followed in the case, but refuse either to examine or criticise the previous attendant's remedies. Let your conversation also refer strictly to the present and future and not to the past, and in no way allude to the physician superseded, unless you can speak clearly to his advantage. As a rule, the less you say about the previous treatment, the better.

To take a mean advantage of any one whom you have superseded, besides being morally wrong, might engender a professional hornet, which, in retaliation, would watch with a malignant eye and sting fiercely wherever opportunity offered. Eschew all sorts of *finesse*, and let courtesy, truth, and justice mark every step in your career. Seek, moreover, to enhance your profession in public esteem on every fitting opportunity, and defend your brethren and your profession, also, when either are unjustly assailed. Indeed, to fail to defend the reputation of an absent professional brother, even by a conspiracy of silence, when justice demands you to speak, is not only unprofessional, but is more or less dishonorable, and implies a quasi-sanction of the libel.

Every physician has his successes, and also his failures. Where you are highly successful in diagnosis, or have worked wonders in treatment after others have failed, observe a proper degree of modesty, and avoid pushing your triumph so far as to wound the feelings or mortify the pride of your less-fortunate predecessors, on the principle of

"Hit him again, he has no friends."

Take just credit, but be guarded in your words and actions, and take no undue advantage of their errors, that you may not in turn invite disparagement or arouse hatred.

"No man likes to be surpassed by men of his own level."

We all know there are a thousand unwritten ways to show an ethical spirit and a thousand undefinable ways to evince an unethical one. When you doubt whether this or that patient is fairly yours or another's, give your rival the benefit of that doubt. Never be tenacious of doubtful rights, but let your every-day conduct, in this and all other respects, entitle you to the esteem of your medical neighbors.

Also, while alive to your own interests, do not captiously

follow up every trifling ethical infringement, difficulty, or apparent contradiction, as if you were ever on the watch for provocations and angry collision with your neighbors, and courted a war with everybody for what you may be pleased to call your "rights." A certain amount of jarring and clashing in a profession like ours is unavoidable; allow liberally for this; school your feelings; bury pettiness, captiousness, and narrowness in the ocean of oblivion, and maintain a friendly attitude toward all fairly-disposed neighboring physicians. Unless you do so, many questions will arise that cannot well be adjusted by an appeal to the code, and you will become involved in useless, rancorous, and endless controversies and reprisals with those whose paths may happen to cross your own.

Sometimes—
"The very silliest thing in life
Creates the most material strife."

You will find it both inconvenient and embarrassing to pass and repass a medical neighbor between whom and yourself there exists a chronic feud, or individual estrangement, jealousy, and hatred, as, also, to meet any one else with whom, through enmity or other cause, friendship and speaking acquaintance have ceased. If ever you have cause to believe a medical neighbor has treated you unfairly, or misconstrued your own conduct or motive, instead of the fierce onslaught and bitter rejoinder, go or send directly to him, and in an earnest but urbane manner make or ask an explanation.

Eschew all doubtful expedients that relate to getting patients and profits, as though you cast off or assume the code of ethics just as suits your purpose; and be very careful not unjustly to encroach on any other physician's practice; also, never attempt unjustly to retain any patient to whom you are called in an emergency; if you are in doubt whether you were deliberately chosen, or only taken in the emergency, do not hide yourself behind a mean technicality of ethics, but ask the direct question. If you learn that another was really preferred to you, surrender the patient to him on his arrival, even though you may be, for politeness' sake, asked to continue in attendance. Circumstances may even require you to have the former attendant sent for in a case, either to take charge of it or for consultation.

Acts of neighborly kindness are frequently performed by physicians for one another, and go far, very far, toward neutralizing the ruffles, stings, and collision of interests which the very nature of our profession makes inevitable. If your conduct toward other physicians at such times is invariably just and honorable, as if arising from a simple desire to do that only which is right, it will in due time be recognized and appreciated, and will not only assist in making your road pleasant, but, if you ever unwittingly infringe, one and all will acquit you of any intentional error.

When you are called, in an emergency, to prescribe for a patient who is under the care of another physician, it is better to leave for him a copy of your prescription, that he, knowing its exact character, may be able to judge whether or not he should continue its use.

Be it your invariable rule never to visit a patient who is under the care of a brother physician, as a "smelling committee," or medical detective for the patient's beneficial society, with a view to ascertain whether he is malingering, or for an employer, friend, or relative who is anxious and apprehensive in regard to his illness, or for one in fear of an impending damage-suit, with a view to report thereon, without the distinct sanction of the attending physician. It would be a still greater wrong to clandestinely remove the bandages from fractures, ulcers, etc., applied by another physician, whether it be to change treatment or merely to examine the case.

Be also extremely discreet and chary of visiting patients under the care and treatment of other physicians, even for social purposes, as it is a frequent cause of suspicion and contention.

Never take charge of a patient recently under the care of **any** regular physician without first ascertaining that he has been formally notified of the change. The principle that governs such cases is this: When a person is taken ill he is at liberty to select any physician he prefers, but after making a selection, and when the case has been taken charge of, if for any reason whatever the patient wants to change, he must, in doing so, follow the established custom, for if there are any hard thoughts against the other physician, or unpleasant scenes with him, the patient and his friends should have them, not you.

The dissatisfied persons who wish to discard their medical attendant and employ you, will sometimes contend that the rules relative to taking charge of patients, recently under the care of another physician, are harsh and unjust, and peculiar to the medical profession. Neither of these statements is true, for our custom is identical with that which prevails everywhere among all classes of people, which requires the formal discharge of the old employé before a new one can take his place. Besides, no person, whether menial, mechanic, or physician, can fill a vacancy till one exists.

Be especially chary of taking cases in families into which you have ever been called in consultation, more particularly if you were called in at the former attendant's suggestion, on account of your supposed greater merits, for he, chagrined at his displacement, will be apt to scan every feature of the change, and, if there be any ground at all for suspicion, he will conclude that, instead of obeying the Golden Rule, and sternly refusing to supplant him, you have taken advantage of the introduction *he* gave you, ingratiated yourself in, and ungenerously elbowed him out.

"I taught you to swim, and now you would drown me."

You will sometimes be called to a patient, and, upon going, will find that he is under the care of some other physician, and will, of course, refuse to attend; but you will almost surely be urged just to look at the patient and tell what you think; or whether the attending physician's treatment is not wrong; or to prescribe for him; with the assurance that the other physician shall be kept in ignorance of your visit. Bear in mind that honor and duty require you to do right in these and all other positions in which you may be placed; not through fear, or for policy's sake, but because it is right to do right, and for the other equally broad reason that you yourself would be cognizant of the wrong, whether the other knew of it or not, and it would lower you in your own eyes; decline, therefore, courteously but firmly, their solicitations, with an impressive assurance that you desire to possess your own respect as earnestly as you do that of others. Unless a great emergency exists, you should determinedly refuse either to sit in judgment on another's work, or in any way to interfere; if, however, the case be one of urgency, your services should be rendered for the attending physician, and you should leave a note telling him what you have done. Take care to make no charges for such services.

When persons are inveighing to you against an attending physician, or one who has been discarded, and finding fault with his treatment, or at the patient's being so long unrelieved, you should never suggest that he be discharged, so that you may supplant him, as it would seem like piracy, or intriguing for a brother's place not vacant.

The rules regarding previous attendance are much less stringent in floating office practice than in regular family practice, and it is not essential to inquire whether an office patient is under the care of another; I believe that all of the most eminent physicians prescribe for all ordinary office patients with but little regard as to who has been attending, or where, or when. Most people, with long-standing, or peculiar, or indefinite ailments, are unwilling to resign themselves to the stroke of Providence until numerous physicians have been tried in vain; and a patient with heart trouble, cough, or a skin disease, will occasionally consult almost a dozen physicians at their offices in as many weeks. The principle followed is simply this: Office advice to strangers is everywhere cash, and the payment of the fee frees the patient to go subsequently to whomsoever else he pleases.

You will see much to condemn in regard to ethics, both in the profession and in the laity. Should you ever feel constrained to attack or impugn any one's conduct, do it in an open, manly way, and never covertly or anonymously,—for underhand, clandestine, and dark-lantern attacks are despicable.

"All ambushed attacks are both cowardly and mean."

Be punctilious in your endeavors to do every person justice. If you err at all in this respect, let it be in liberality. Suffer injustice, rather than participate in it. Sometimes, even though the letter of ethics allows you to take a patient, it may be unkind or unwise, or brutal to do so; use such opportunities to harmonize, rather than to disrupt. You can do this, and yet not make a habit of cheating yourself out of patients.

\* \* \* \* \* \* \*

Always keep some good vaccine virus on hand, both for the fees it secures, when there is a demand for vaccination, and for fear of a sudden outbreak of small-pox.

Vaccination, although a trifling operation, is a prolific cause of criticism and reproach to physicians; take your time, and do it skillfully and thoroughly. In lieu of humanized virus or arm-to-arm vaccination, use calf-virus whenever it is possible to obtain it; it is more popular, and not capable of communicating syphilis, scrofula, etc., and needs less defense. In no case use any but pure virus, and be ever ready to defend its purity with proof if any one you vaccinate suffers any mishap through it.

Remember that you are legally as well as morally bound to vaccinate a person after promising to do so. Besides the regrets and harsh criticism your neglect would generate, a suit for damages might follow, if the patient should get small-pox while awaiting the fulfillment of your promise.

Do not begin the unjust custom of vaccinating children gratuitously, in cases where you have officiated at their birth, as is the habit with some. Make the same charge also for re-vaccinating any one, to test whether his former vaccination is still protective, whether it takes or not, as you would if he never had been vaccinated before, as revaccination succeeds in but a small proportion of those it is tried upon, and the charge is for making the test.

A public vaccine physician should never insist upon vaccinating a child or other unvaccinated person who is known to have a discreet, watchful medical attendant, unless small-pox is actually prevailing. They should, on the contrary, be referred to him.

You should, of course, make no extra charge for repeating primary vaccinations till they take, no matter how long the interval between the trials; also, make but one charge for any person who has revaccination attempted, no matter how often, if during the same epidemic or small-pox scare. Many people believe a vaccination protects as long as the scar shows plainly. The truth is, a vaccine scar lasts for life, while the protective influence of vaccination gradually disappears in some people. A typical vaccine scar merely shows the vaccination once took properly, not that it still protects.

Some people think a revaccination must be made to take anyhow, even though they are still protected by the old one. You cannot catch fish where there are none, no matter how you bait your hook; nor set a pile of stones on fire, no matter how good the matches.

Another error regarding small-pox: Many people imagine that it can only thrive when the weather is cold; this is a mistake, as it may prevail with intensity at any season. Indeed, severe epidemics of it often prevail in tropical countries where there is perpetual summer.

Avoid volunteer practice, and be very cautious how you go out of your way to persuade people to let you remove warts, extract tumors, efface tattoo-marks, destroy nævi or superfluous or disfiguring hairs, and do other minor surgical operations gratuitously, with assurances of success. There is always a possibility of serious or fatal sequelæ; the most trivial operation—

even a puncture on the tip of the finger by a pin, needle, or splinter—is occasionally followed by death, and you should not, especially in private practice, induce people to let you involve yourself for their benefit, without being paid for your risk and responsibility; for instance, it is an ugly matter to have a wart you have insisted upon tampering with become an ulcerating epithelioma. It is better, indeed, to avoid all unrequited work, and all gratuitous responsibility, other than what charity calls for.

For similar reasons do not persuade people to effect insurance on their lives, or in any particular company, as all such ventures carry a possibility of disappointment or failure that might involve you.

Wisdom in recognizing cases that are likely to involve you in suits for malpractice, and in foreseeing and forestalling the suits themselves, is also a valuable power. Take care that this wisdom does not come too late or cost you too much. Remember that when you are employed professionally you are regarded as contracting that you possess and will exercise ordinary skill in your profession, and that you will be guilty of no negligence. Beyond this you are not responsible for the result, no matter how bad, as medicine is not an exact science; but if you fail either in ordinary skill or care, you are legally liable to the injured person to the full extent of the damage sustained. Skill should, of course, be measured by the time and place in which it is exercised; whether on land or on ship-board, in places where facilities are few, or where they are many, are matters to be taken into account. In your professional rounds you will not find the various diseases as clearly marked as they are in the books,—not labeled as plainly as the bottles in a pharmacy,—therefore a mistake in diagnosis is not sufficient cause for action, and every physician may be, and often is, mistaken; indeed, many cases are so obscure, or masked, or irregular, or complicated, that nothing but an autopsy, and sometimes not even that, can reveal their exact nature.

Never fail promptly to send in your professional account to dissatisfied patients who may be unjustly attempting to injure your reputation and practice, and especially to such as may be threatening to sue you for malpractice, whether or not you expect them ever to pay it. If you cowardly shrink from doing so in such cases, it will be quoted as proof that you are guilty of what they charge, and that you know it. The presentation of your bill will give you a better position before the public, and raise an issue that greatly tends to checkmate them. In all such cases *do not fail to charge the maximum fee.*

When you are to be a witness in court in a grave case, courteously but firmly decline to give any person connected with the opposite side either a verbal or written statement of what you saw, heard, or observed in the case, or what your opinion is, or what your testimony will be. Also, if need be, dispute their right to question you at all on the subject.

If you are yielding in this respect, you may actually aid them to set traps for you, by distorting your statement from its proper meaning and intent, or to rebut it on the witness-stand, or to prepare to charge that you are lacking in medical knowledge, and thus bring both justice and yourself to grief. Often, in such cases,

"Your enemy makes you wise."

Firmly but courteously inform such agents that you will not give the desired information, but that they can elicit all you know on the witness-stand.

When giving evidence in court, whether as plaintiff, defendant, or witness, endeavor to keep cool and self-possessed, and give your evidence with manly and honest candor; guess at nothing, and express no opinion for which you cannot give the why and wherefore.

There is no class or profession other than our own whose members habitually confront and confute one another in the courts and before the public. Our so-called psychological experts, specialists, and other would-be highly scientific representatives, have so often been hired by contestants with a view to use their dialectic powers to frame or elicit favorable testimony, or the reverse, as the case may require, in will, life-insurance, criminal cases, etc., that the public are led to freely jest about the differing opinions of physicians, and not unnaturally to believe, from our lamentable professional contradictions and divergence of opinion, that there is no case so disreputable, no claim so monstrous, that it cannot be bolstered up by medical evidence; and that our boasted science of medicine is merely a tissue of guess-work, and that a certain class of pseudo-experts can make things appear to be either black, white, or lead-colored, and are willing to sell testimony to the highest bidder, on any side of any question.

So-called "medical experts" often excite disgust and indignation at the contemptible attitudes they assume when they act against their better knowledge, and join hands with mercenary and venal people to attempt to mulct a physician, or to free a criminal from legal responsibility, perhaps to let go a murderer whom all the world knows is guilty, or to condone other scoundrelism on the plea of "insanity," "hypnotism," or immoral pretext gotten up to make money, defeat justice, or obtain notoriety—

"And help to blind both judge and jury, not to give them eyes."

Never forget that every principle of honor and duty requires us to stand by and defend each other in everything that is reasonable and just, and forbids us to think of lending ourself as "medical cat's-paws," either to go on the witness-stand or to prompt council in their efforts to bandy and break down medical witnesses on cross-examination, in rascally or speculative malpractice suits against reputable physicians who have conscientiously discharged their duty in cases of sickness, accident, or surgical operation. Fractures about the wrist or elbow furnish a large proportion of these cases; eye cases, also, furnish another large share.

These slanderous suits against physicians are generally trumped up and entered either at the instance of designing physicians intent on the ruin of rival practitioners, or by unprincipled, case-hunting, Champerty lawyers,—

> "The words of their mouths are smoother than butter,
> But guile is in their hearts,"—

not with the hope that they may come to trial on their merits, but that the accused physician, through natural dread of the expense and annoyance, will pay a snug sum as *hush* money.

The court records make it appear that the poorer a patient, and the more that charity has been exercised, the more likely he is to enter suit and otherwise show the basest ingratitude. If ever a worthless, lying loafer gets a chance at your pocket-book, look out for him.

Probably there is no department of professional duty in which physicians are asked to *stretch* their consciences so much as that of giving certificates that the disability of persons seeking to get soldiers' invalid pensions was contracted in the army.

It is also possible that you may be cajoled by friends, or blandished or flattered by interested strangers, or even tempted by gold, to give an opinion that old Jinglecash, who was mentally unfit to make a will, was unclouded in mind and fully competent to do so, or that Mr. Drinkhard or Mrs. Halfgone, with one foot in the grave, the result of intemperance or disease, is sound or temperate, and thereby to swindle an insurance company; or that Mr. Badbody or Mrs. Dysoon, with a bias toward a certain disease or with an incipient organic affection, is in perfect health. Or Highflyer or other pleasure-loving officials may seek to cover their absence from duty by your certificate that their non-attendance was due to sickness; or Mr. Makout may attempt through your aid to escape military or jury duty, or attendance at court as a witness, or for trial, or try to get from you a prescription for a "Sunday drink of liquor" for the thirsty, under the old pretense of "very sick."

Repel all such attempts promptly and decidedly, and emphatically refuse to be seduced from the path of honor and integrity, or to deviate from your honest conviction, for any one.

With professional honesty for your pilot, be firm and unwavering in your determination to steer clear of practices and alliances in which your part would not bear legal scrutiny or detailing in the community; and you will not only safely pass the various rocks of shame and whirlpools of bitterness which have wrecked so many of our profession, but you will have the full approval of your own conscience. Perish all that conflicts with the attainment of this.

## CHAPTER III.

*"Whatsoever a man soweth, that shall he also reap."*

WHEN you are importuned to produce abortion, on the plea of hiding from the world the yet-undiscovered guilt and saving the poor girl's character, or preventing her sister's heart from being broken, or her father from discovering her misfortune and committing murder or suicide, or him who has taken criminal advantage of her from being (*sic*) disgraced, or to avert the shame that would fall on the family, or the church scandal about one of the weak brethren; or, in cases where there is no previous guilt, to limit the number of children for married people who already have as many as they want, or who are just married and do not want the inconvenience of them so soon, or to accommodate ladies who assert that they are too sickly to have children, or that their suckling child is too young to be weaned, or that they have been pregnant only a short time, or to avoid other anticipated evils, etc., etc., even though it be only the size of a mustard seed, you should not stop to discuss the subject lengthily with a "h'm" and a "haw," but should meet all such entreaties and solicitations with a refusal prompt, strong, and positive, and never let yourself appear to entertain the proposition. If they are too importunate, express your sentiments in unmistakable language, and with plain, American frankness, bow them out, but remember that these are terrible secrets, and seal your lips doubly tight.

It is always safe to do right, and never safe to do wrong. How could any one but an idiot, or an utterly unprincipled man, be induced to stain his hands and his heart by committing a crimson crime; to violate both his moral conscience and the criminal law; to risk exposure, social disgrace, and professional ruin for himself and family, and even the penitentiary itself, by taking the guilty burden from others' shoulders to his own, thereby putting himself in their sinful power, whether as a favor or for a paltry bribe, or even for all the gold of California!

Evil rumors fly rapidly. The production of a very few criminal abortions (sometimes even a single one) will surely go from tongue to tongue, and give the damphool physician who stoops to commit them a widespread notoriety as infamous and as tenacious as the Bloody Shirt of Nessus. Take care

> "That the immaculate whiteness of your fame
> Shall ne'er be sullied with one taint or spot."

A single misstep from the heights of integrity may wreck one's whole life.

When circumstances render it necessary for you to prescribe for females with suspended menses, where pregnancy is possibly or probably the cause, it is better, instead of giving a Latinized prescription, to order some simple thing, such as hop-tea, tincture of valerian, or wine of iron, under its common English name, with full written instructions how to take it. By thus avoiding all secrecy regarding the nature of the remedies prescribed, you will avert the suspicion or, may be, a charge of giving abortifacients.

To give a woman who applies for an abortifacient an inert agent would, to say the least, be unwise; it is better plainly to refuse to give her anything, whether a pretended or real remedy.

The charge or suspicion of criminal abortion is much more apt to be brought when the woman is single than when she is married.

> "The physician must, like the diplomatist, tread softly."

You must give a cautious, a very cautious opinion, if any, in cases of unmarried females whose menses have ceased and pregnancy is feared, or as to whether an apparent pregnancy is real, especially in a case where the suspected girl, after everybody else has left the room, strenuously denies having had carnal intercourse. Many will not confess the truth while a third person is present. Erroneously to pronounce an honest, virtuous woman pregnant may blast the whole future life, honor, and innocence of one who was provided with a shield of virtue and clothed in the mantle of purity,—

"A soul as white as heaven,"—

and call down maledictions on you; if, on the other hand, and on insufficient evidence, you too quickly declare her "not pregnant," or that it is "the dropsy," or "a tumor," it might seriously injure you; but this mistake would bear no comparison to the former, or to the injury you might inflict on an innocent person by an inconsiderate and fallacious opinion. In every instance, therefore, in which the slightest reasonable doubt exists, temporize or suspend your opinion for weeks, or even months if need be, till positively certain that it is "a kicking tumor," by hearing the fœtal heart-beat or feeling the fluttering of the child within the uterus, or some other unequivocal sign.

Unmarried negresses, ladies of easy virtue, and other low females (and sometimes even the wealthy, young, and beautiful; in silk, satin, velvet, and gold), who fear they are pregnant, will occasionally come to consult you, consume your time, and get your opinion, and when you discover that they are really pregnant, and refuse to produce abortion, will try to escape the payment of your office fee. Where you fear such injustice, courteously inform them at the beginning how much your fee is for your time, opinion, and advice, and that it must be paid whether your recommendation agrees with their wishes or not. After settling the fee question, study their case, and candidly give them your opinion and advice.

Should you ever encounter a case in which you believe the destruction of the unborn child is (for physical reasons) necessary to save the mother's life, do not consent to do it secretly, but only after regular consultation with some other physician of well-known probity.

To give directions for the prevention of conception; or instructing in onanism, buggery, or other nasty conjugal sins; or in the guilty use of condoms, sponges, syringes, or preventives against venereal diseases, that encourage the timid to venture;

or in other instruments or expedients to aid crime or to defeat nature; though offenses beyond the reach of the laws, is, nevertheless, most derogatory and degrading to the physician, and a disgraceful violation of his professional office.

Never carry away or keep chloroform, ergot, splints, instruments, or other unused articles that patients have paid for, without a clear agreement with them to that effect; and be very, very careful how you infringe upon the wine or liquor intended for a sick person, or eat his cake, fruit, etc. Foolishly to do such things would not only lay you open to criticism, but even to the most mortifying charges of meanness or dishonesty if a rupture of friendship should ever occur,—in fact, with such things to fortify them, many people would be somewhat disposed to welcome or create a rupture with you.

Be careful that attempts to conceal the presence of contagious diseases, or other recognized sources of danger to health, or of births resulting from clandestine marriage, or from bastardy, do not involve you in the exposures and recriminations that are apt to follow.

If you have skill in avoiding cases likely to render your attendance necessary in court as a witness and other time-consuming annoyances, legal and social, it will prove a source of much comfort and relief.

Cultivate agreeable relations with your professional neighbors and keep old friendships in repair. The practice of medicine isolates the members of our profession from one another much more than one would suppose. Neighboring physicians, fellow-workers in the same humane and beneficent profession, and well known to each other by sight or reputation, daily pass and repass each other without a look or nod; and, although acquaintanceship and social amenities might be mutually agreeable and beneficial, and possibly ripen into life-long friendships, they often remain as strangers for years, unless some fortuitous circumstance brings them together.

Two and two are four,—this is always true, whether we

are counting pebbles, people, or planets, but it is no more true than that every physician ultimately rises or falls to his proper position among his fellows.

> "Pygmies are pygmies though perched on the Alps,
> And pyramids are pyramids in vales."

Determine, therefore, that you will become something more than a mere *visiter* of the sick.

From the very beginning of your career you have social and fraternal duties, as well as individual and solitary ones; hence neither hold yourself aloof from the profession nor attempt to isolate yourself, and attend to your own interests merely; but identify yourself, head and heart, with your medical brethren in all legitimate public professional matters: attend the medical conventions, assemblages of alumni, medical meetings called to provide entertainment for visiting medical celebrities, memorial meetings held to pay special tributes of respect to deceased medical brethren, general meetings of the profession, held to voice the opinions or policy of the profession as a body, regarding public dangers, or to take associated action on matters of public hygiene, or regarding medical laws; or to devise and urge the adoption of sanitary measures against epidemics, etc., etc. Your presence at these unions and reunions will keep you in touch with the profession, and be an earnest of the spirit that actuates you.

Also, join the medical societies of your neighborhood; and if none exist, induce your medical brethren to join you in founding one. Organization gives protection both to the profession and to individuals. Society membership is a guarantee of your good standing and that you pursue legitimate practice.

A good medical society is also something of a post-graduate school.

> "Steel whets steel."

And, next to actual personal experience, there is nothing so valuable to the young practitioner as the medical society, for there the collision of mind with mind, and of thought with thought, in amicable discussion, awakens reflection and deeper reasoning, increases the intellectual grasp, stimulates the mental digestive power, and liberalizes and enlarges the scope of both the speaker and the listener, and acts as leaven to the entire profession. Nowhere else can you study so well the individuality and the styles of different physicians, and discover the reasons why each one is what he is, so fully, as at medical meetings. There the specialist, the teacher, the general practitioner, and the book-worm all meet,

"Well armed with mighty arguments,"

and each in his own way contributes to the instruction and intellectual recreation of the others. There you can meet your neighbors on common ground, grasp each other by the hand, look into one another's faces, and compare investigations, experience, and opinion by face-to-face discussion.

"Many things, obscure to me before, now clear up, and become visible."

There rivalries, dissensions, jealousies, and controversies can be softened, and professional friendships be formed and cemented; there you can find opportunities for pleasant, social intercourse with worthy men. There you can also silently measure the height and depth of your medical contemporaries, and see the difference between the serious and the superficial thinker, the convincing and the faulty logician, the judicious and the injudicious, the alert and the stupid, intellectual giants and mental dwarfs; there you can also estimate the influence of pleasing actions and deportment, and the intellectual and moral worth of those who command respect, and discover and learn to avoid the glaring imperfections of others who do not,—and in many other respects learn effectually to separate the chaff from the wheat.

Medical societies, of course, are neither a specific for all personal deficiencies nor a panacea for all professional sores. There you may find men good enough ordinarily to appear with the best, but weak enough, under temptation, to behave with the worst; some, too, who neglect all their better duties under the plea of "lack of time," and attend only when there is to be an election, a feast, or a quarrel. Spending a few hours among honorable physicians once a week will not lift Prof. Sinbad into angelhood, change Dr. Buffoon into a gentleman, or convert Dr. Trickmore or Dr. Quackfrombirth into professional Chesterfields, or lend Dr. Oilyone or Dr. Doubleways consciences like Milton's. But, to repeat: intercourse at a medical society does serve as an intellectual exchange, where one may hear the discussion of moot points and live questions in medicine, and at the same time establish with his brethren friendly and honorable relations. One often sees distrust converted into friendship merely by acquaintance.

Independently of the benefits and improvement accruing to the members of medical societies individually, they give a sound and healthy tone to the entire profession, stimulate the growth of medical science, and also generate and keep alive a genuine professional and brotherly spirit that tends to minimize all that is unprofessional.

Never oppose the admission of any clean-handed, honorable, and competent person into society membership for private or personal reasons, or for any cause other than ineligibility or unfitness for the honors and benefits membership confers, because such societies exist for the advancement of medical and surgical knowledge and for the benefit of all regular physicians, and it would be unjust to mix private feelings with professional duties, and interpose an objection or a blackball on purely personal grounds.

Do not hesitate to take part in the medical debates whenever you have anything valuable to offer, whether it is gleaned from literature, or from the great school of experience. If your views differ from another's, express them with courtesy and respect. If you have a contribution or new fact to offer, an invention, or new pathological views, or a discovery or new secret to announce, a new instrument to show, an operation to describe, a patient or specimen to present, a report to make, or a new treatment, a new therapeutic agent, a promising theory or a talismanic charm to tell of, or anything whatever to say, do it in a careful, clear, methodical manner, then sit down; but when you have nothing worth offering, do not talk for talk's sake, but make Ciceronian silence your law, and do not break it. When on the floor, take care neither to abandon your medical vocabulary for the vernacular, nor let your professional manner degenerate. This will soon teach you to arrange your thoughts quickly and to express them clearly.

Remember in debate, as elsewhere, that there is nothing infallible; that the physician must school his prejudices and be open to conviction. Toleration of a difference of opinion is a lofty virtue; therefore, say or do nothing to wound the pride or feelings of any other member, and if any incautious remark, misstatement, or personal reflection drops from your lips, be not slow to make proper atonement. Those who, Nero-like, are always positively right, while all others are positively wrong; who can brook no opinion that does not accord with their own, are usually deemed hot-headed, rash, and indiscreet, and very unsafe guides. Also, remember that differences of opinion are quite compatible with friendship, and that controversies, discussions, and parliamentary battles, no matter how sharp or excited, are usually conducted by men of discretion within the bounds of decorum, and without violations of the ordinary rules of good breeding; and, also, that there is no mode of practice nor remedy for any disease which has not been the subject of obstinate dispute, and that every great discovery or startling announcement stirs the whole medical world to testing and reporting, asserting and denying.

You will find that many people entertain a belief that medical societies exist for the pecuniary advancement of their members, just as trades-unions and like organizations strive for fewer hours and more pay for the working-classes, and that, in some way or other, they tend to limit the freedom of personal opinion and abridge the individual rights of their members. Be careful to correct such errors on all suitable occasions, and to inform those thus misled that medical societies exist not for selfish, but mainly for scientific, purposes, and the public good.

Keep up your medical studies, or the knowledge which you have already acquired will soon become misty and ere long slip from your memory. Without more or less continuous study the details of cases and the symptoms of many diseases are apt to be forgotten; indeed, after two or three years have elapsed, the mind does not often bring back the details of parallel cases, or of cases for comparison, unless they are extremely uncommon or interesting, and their utilization is thus lost to mankind. Test your memory now by asking yourself the following questions: What did you have for breakfast on the third day of last month? What kind of a day was the ninth of last February?

In consulting journals and text-books, remember that practice found successful in your own climate or region is, as a rule, more to be relied upon locally than that applicable to the same disease in other climates. Also, avoid relying on antiquated works on practice and back volumes of journals as guides in so progressive a science as medicine. New investigations and rapid progress render new text-books essential to those who would keep up with the medical world and maintain the skilled readiness and self-reliance which the consciousness of being fully posted on new instruments, methods, and improvements naturally inspires.

Endeavor to collect and form a library of standard professional works as soon as possible after graduating; books are the tools, the literary apparatus with which we cut and dig our way to knowledge, and we now have more books and better books than ever before. Money spent in this way will return a hundredfold. There is an art in selecting material to read; buy the best authors and always the latest editions, but take care that irrepressible book-agents, with "the greatest work ever published," do not induce you by their importunities to subscribe for a jumble of books for which you have but little or no use. No one can patronize everything, or even read one-tenth of all that is offered, unless he has nothing else to do. You need not be ashamed of a library of twenty or thirty well-selected volumes of recent date, provided you have thumbed them well, and are familiar with their contents; and were you even to buy one volume at a time and study that well before getting the next it would be no mistake.

Subscribe to one or more medical journals and scientific publications, and read and digest them carefully, so as to keep abreast of the discoveries and theories of the passing day. They are necessary to the progressive physician. But neither swear at nor by all you see in them; be especially distrustful of publications, edited by Dr. Inkpot or Prof. Penn, that exist for the purpose of advertising either their owner's hobby or his goods, or a college or its clique. As a rule, you will find that statements found in the text-books and in standard monographs are more mature, more pointed, and more representative of collective learning, and, in relation to therapeutics, generally much more reliable than those in journals, which are often founded on a single case, or the fine-wrought theories or exaggerated fancy of some unbalanced rainbow chaser,

"Educated beyond his intellect;"

or the unconfirmed experience, representations, expectations, or speculations of some partial observer, riding a hobby or pitting himself against everybody.

Take care to have a good Dispensatory and a work on Medical Jurisprudence among your books.

Acquaint yourself fully with the contents of your library, so as to be able to refer to whatever you need without hunting; also, have one certain place for every book.

Never allow yourself to be biased too quickly or strongly in favor of new or unsettled theories, based on physiological, microscopical, chemical, or other experiments, especially when offered by the overzealous to establish their own conclusions or preconceived ideas, or by those who have identified themselves with the latest medical novelty.

Also, do not allow yourself to be led too far from the practical branches of your profession into histology, pathology, microscopic anatomy, refined diagnostics, bacteriomania,—

"Ha! ha! thou, too, hast some crotchets in thy head,"—

comparative anatomy, biology, psychology, the arrangements of electrical currents in muscular fibre, and analogous subjects, that merely interest or create a fondness for the marvelous; else it will impair your practical tendency and give your mind a wrong bias, and your usefulness as a practicing physician will almost surely diminish. The first question for you, as a practitioner, to ask yourself in everything of this kind is, What is its use?

I would not apply these remarks to school-men, or to professional teachers and experimenters, who have hospital and laboratory facilities, and, perhaps, wealth and leisure, and are nobly pursuing the higher reaches of purely scientific investigation and original thinking on borderland questions, chiefly for unselfish love of them, or to gain FAME or distinction therein, and become truly great;—

"That man is great, and he alone,
Who serves a greatness not his own;"—

or to others who, being favorably situated, are delving solely for the pursuit of pleasure, and not looking to their practice for support. Nor would I dare say these are not priceless kinds of knowledge. I mean to say that skill in the practice of medicine does not depend so much on what the practitioner knows abstractly as what he knows and has the use of, and that a person may get so deeply absorbed in the hemi-, demi-, semiquavers of the deep labyrinths and fine subtleties of science and high-thinking as to regard nothing but them, and that your most useful studies as an every-day practitioner will be the well-ascertained facts of the profession, which are essential to every skillful physician: knowledge that relates to the structures connected with accidents, operations, and surgical affections, and to those of the organs that are the principal seats of medical diseases; practical subjects required for the daily duties of the profession, and, above all else, the art of treating diseases with success. To know how to relieve a colic, pass a catheter, or cure a node, is a thousand times more valuable than to know that the anterior cornu of the fourth ventricle of the brain runs a course that is backward, outward, downward, forward, and inward!

The great popular test of medical skill is curing the sick; and you will find that your reputation will depend more on the successful treatment of your cases than upon familiarity with the ultra-scientific; and you will meet physicians, possessed of comparatively small knowledge, so dextrous in its use that they have done great good in the world, and ridden over the heads of some far better versed in the books.

Never, for the sake of appearing in print, publish trifling or hastily prepared medical articles, as whatever one writes is naturally supposed to be a mirror of his own mind. Do not, however, hesitate to write whenever you have anything valuable or instructive to offer, both for the benefit of others and to enhance your own value, reputation, personal respect, and dignity.

"Of all the arts in which the wise excel,
Nature's chief masterpiece is, writing well."

All people respect the man who thinks.

When possible, base your articles on solid facts, or on an analysis of facts, rather than on speculation and theory. Let your diction be pure and simple, and as short and aphoristic as perspicuity will allow, so as not to weaken the effect of your ideas, or obscure them in a lot of long-winded or idle verbiage; rather go straight to the point, and make every word count, in expressing clear, bright ideas, and let accuracy be characteristic of all you write.

Be especially careful to give your paper, essay, or book a concise, appropriate, and, as far as possible, an attractive title,— one that indicates its contents, and shows with sufficient clearness the general character, purpose, and point of the remarks which are to follow.

> "Oh, how that title befits my composition!"

This is essentially necessary when the title of the work is to be put in an index or catalogue. Such indefinite titles as "A Curious Case," "Clinical Communication," "Plain Facts,"—

> "Bless us! what a word on
> A title-page is this!"—

"A New Method," "A Case of Interest," etc., furnish no clue whatever.

In writing, cultivate perspicuity, precision, simplicity, and method; avoid flaws of grammar or logic, and unmerciful diffuseness, and do not interlard with far-fetched, jaw-breaking scraps and patches from the dead or foreign languages, unless a translation be appended; for, unless it be some time-honored phrase, or hackneyed quotation, the average reader will probably be forced either to pass it over unsolved, or take down his classical dictionaries, dusty book of quotations, or his school-boy grammar; besides:—

> "Every man is not bred at a 'Varsity.'"

The English language, the language of Shakespeare, Milton, and Bacon, is of itself capable of giving lucid, eloquent expression to every thought of man, and it is to be regretted that Fortislingua, or any one else, should, from superfluous wisdom, or pedantic pretension (Anglographic aphasia), fail to express himself in his own mother-tongue, and make his work brilliantly incomprehensible by throwing in handfuls of Latin and Greek, almost as a cook peppers his broth, as if

> "This writer has been to a great feast of languages and stolen all the scraps."

The recent attempt to supersede the old weights and measures (which every one understands) by the foreign-looking metric system did not succeed; it is therefore scarcely worth while now to discuss its merits. When you report cases or publish anything in which weights are given, either use the familiar English weights and measures, or give both the old and the metric; to use the French system only savors of affectation. The average reader makes no attempt to carry the metric equivalents in his mind, and if you give metric measures only he may not take the trouble to calculate, but pass your effusions by without getting the information you wish to convey.

Take notes of all remarkable cases, but do not report or publish any that are not unique, or at least that do not present some curious, rare, or very instructive feature, or militate in some way against accepted theories; otherwise, you will merely increase without adding anything valuable to existing records. You will find every department of medical literature is fast becoming loaded down with theoretical discussions, speculative dissertations, compilations, and word-building; old, universally-known things said in a new form; many

> "An anxious blockhead ignorantly read,
> With loads of learned timber in his head,"

seeming to say:—

> "In pity spare me, while I do my best
> To make as much *waste-paper* as the rest."

You should omit book-matter generally known, and contribute original work, new things rather than new phrases, new ideas rather than new words. Use a plain, intelligible style; do not count your words, but see that every word counts; also, avoid such ambiguous descriptions as "the color of an orange," "the size of a strawberry," "about three inches long," "about as thick as blood," etc.; and be as brief and concise as justice to your subject will allow, and, for the poor printer's sake, prepare your matter so as to please his eye and require but little or no revision on account of grammatical errors, bad phraseology, or faulty style of construction.

When you begin authorship and write books, essays, or monographs, use, for the sake of convenience, the smallest-sized sheets of white note-paper, and avoid rolling; this will enable you to keep them flat and to handle them more easily in writing, altering, and rewriting pages; also, to carry them to and fro, and to preserve them much better than if large. If intended for the press, write only on one side of the sheet, and leave a margin at the edge.

Be careful, also, to avoid the useless custom of appending to your name an excessively long list (like the tail of a comet) of all the titles and alphabetical appendages that you can rake together, with half a dozen etceteras; such enumeration is in bad taste, and tends to excite the ridicule of persons of discernment. The chief use of suffixes is that the identity of the writer may be recognized; a single suffix, or simple title, or the name of your town, street and number, are unpretentious and yet sufficiently explicit. Some publishing houses evidently think the use of titles by authors who have reputation as writers aids the sale of a work.

Never furnish a report, statement, or opinion on any important case or subject for publication, either in book, journal, or newspaper, without a proviso that you are to see, and if necessary revise the proof, and correct the printer's errors in spelling, punctuation, etc., before it goes to press; otherwise, you may find some purblind proof-reader or go-ahead printer making you say the reverse of what you intended, thus necessitating a long list of "errata," or may be causing you to regret that you ever allowed the article to appear in print.

Do not fail to pay your honest debts punctually, even though you be cheated out of half you earn. The best plan is to restrict your expenditures to your income, and pay as you go, and if you cannot pay much do not go far; for to be in debt for horses, carriages, horse-feed, or, still worse, for dress, luxuries, rent, servants' wages, etc., cannot fail to set the tongue of scandal to wagging freely and injuriously, to the possible ruin of your credit. Payment must be made sooner or later, and it is far better to discharge each liability as it becomes due than to be paying those that should have been paid a month or a year ago. Be especially careful to keep your medical society and journal dues paid promptly, and to discharge all other pecuniary obligations. To borrow books, instruments, umbrellas, money, etc., especially if you keep them beyond the proper time, or return them in bad condition, will also tend to depreciate you more with the lenders than you would suppose. Never involve yourself by borrowing any apparatus, instruments, etc., from one physician, or patient, to lend to another; if necessary, introduce the parties to each other, and let the borrower borrow on his own responsibility.

It is needless to say that health and decency require you to guard against uncouth, untidy, and repulsive habits; do not pick your teeth or pare your finger-nails, or squirt tobacco-juice around you at your visits, or have your breath. hair, and clothes as redolent as a bar-room spit-box with pipe or cigar fumes, alcohol, stale tobacco, dead beer, etc., or with cloves, cardamom, and other masking aromatics, or the smell of iodoform, carbolic acid, and other disgusting medicines on hands or clothes, or you will unavoidably prove obnoxious and disgusting, and invite criticism and possibly engender aversion, and entail the loss of your patient. Coarseness and vulgarity are sufficiently disgusting in anybody and under any circumstances, but in a physician, and especially in the presence of females, they are unpardonable.

Avoid every habit that can give reasonable offense: to make your appearance in your shirt-sleeves, with unwashed hands, dirty finger-nails, dingy cuffs, egg-spotted or tobacco-stained shirt-bosom; greasy coat, out at elbows; ragged pants, fly-speckled or crumpled hat, shaggy whiskers, or four or five days' beard on the face; rough, creaking, or dirty boots; or with pipe or stump of cigar in mouth, or chewing a quid of tobacco; or skylarking, showing unseasonable jocularity; using coarse, vulgar, and impassioned language; habitual swearing, loud guffaws, etc., will by many be regarded as evincing moral weakness, and tend to diminish your influence and prestige, detract from your dignity, and greatly lessen you in public esteem, by impressing people with the idea that, after all, you are but an ordinary person, and not up to their ideal standard.

Moreover, to be seen carpentering, painting, or displaying other common-place or out-of-place talents, would also suggest that your mind was not engrossed with your profession. You may possibly secure faith in spite of these, but usually such proclivities unquestionably tend to decrease it.

The nerves and tactile corpuscles of the tips of your fingers will have much to do with your skill and success; these nerves are sometimes even superior to the sense of sight; to palpate the chest or abdomen, examine tumors, make vaginal examinations, do surgical work, etc., the hand must be steady and the touch must be nice and delicate. If your fingers, instead of having their sensibility protected and their tips educated, are rendered callous and clumsy by manual labor or rough usage, their delicate nerves will be unfit for these duties.

Beware of a certain temptation to which the practice of medicine especially exposes you. The irregularities, anxieties, and exhaustions; the cold, the wet, the hunger, the night-work and loss of sleep, and the hospitality of patients and other friends, all unite to tempt physicians to use alcoholic stimulants. Remember that, although drunkenness and the idle life associated with it may be tolerated in physicians who are fully established in practice, because confidence and friendships had been formed and their talents and worth had become known previous to the formation of the habit, it would be fatal to any one in the formative stage of reputation, or just beginning to gain the confidence of the community; for no one who begins life burdened with this vice will be trusted or employed. Even when the older physician, who drinks, is employed, it is done with loathing, and only to make use of the good half of him, which cannot be separated from the bad, and his visits are looked for by those whom necessity puts into his hands with disquietude and dread.

What is a more disgusting spectacle than a drunken, swearing, reckless sot-of-a physician, with whisky-soaked breath, staggering around the bed of a sick or dying person, profaning the occasion by the thoughts he excites, and by his grossness? The wisest policy for you personally is to avoid intoxicating drinks, which cause so much crime, sickness, and poverty, and allow others to do as they think best. The cause of drunkenness is drinking, and if you are foolish enough to drink liquor, wine, or beer when people offer it to you, you not only run the risk of getting fond of it, but nine chances to one those very people will be the first to add the charge that "he drinks" whenever any other person says anything else against you; but when it becomes known that you never touch the demon it will be of immense advantage to your reputation. But intemperately to urge puritanical ball-and-chain temperance on others, or being an officious member of temperance, secret, or beneficial societies, will aid you but little if any in the acquirement of practice,— the most desirable class of which is the quiet family business that you will attract by a faithful and kindly endeavor to do your very best for all who apply for your services as a physician.

A physician's life, like a pantomime, is full of wonderful changes, and, being a public character, he knows not the day he may need the friendship or good offices of this, that, or the other person toward whom he may have felt, and unwisely shown, political, or religious, or personal hostility; therefore, do not allow yourself to grow morbid on temperance, total abstinence, local option, prohibition, and other sumptuary crusades, partisan strifes, and puerile contentions; as they will be apt to recoil on your head if you make yourself prominent in them. If your office is located very much nearer the church than the tavern, and if you lean to the sabbatarian element instead of the pitfalls of infidelity and atheism, so much the better; but proselyting and pushing matters of a partisan, political, radical, or secular nature is not your function, and you cannot become officious in them and their irritating methods without setting (about) one-half of the community against you, and exciting enmity, and maybe personal hostility. You had, therefore, better leave all subjects for discord, heart-burnings, animosities, and angry discussion, whether political or religious, to the general public, unless your pecuniary or social position is such that you can very well afford to run the risk, or are driven to do so by conscientious scruples that outweigh all other considerations; and even then it is better to let your profession occupy the dominant place and your patients be your first and principal care.

When requested to write a prescription to enable an ailing person, who really needs it, to procure liquor on Sunday or in a local-option district, comply with becoming good nature, but accept no fee for it.

Presents from fond or grateful, very liberal or romantically generous patients, although flattering, will almost invariably lead to the disarrangement, if not actual rupture, of the legitimate pecuniary relations previously existing between yourself and the giver, which it may consequently be impossible fully to re-establish.

"In the long run, gifts are often losses."

Most practitioners can probably recall instances in which presents of knee-blankets, whips, baskets of game or fruit, boxes of cigars, wine, pet animals, canes, free passes, gloves, new hats, curiosities, baby-named-for-you, etc., have spoiled their bill, and proved not only unprofitable, but exceedingly expensive. When you foresee such a result, be guarded.

You will find it a good rule to decline all presents and favors that are likely to place you under embarrassing obligations to patients. A still more important rule is to avoid mixed dealings and crossed accounts with careless hucksters, grocers, feed-men, milk-men, and other patients, as such dealing will rarely continue to be satisfactory; they often lead to disagreements, and in "squaring-up" will almost always result in your getting only about half as much for your services as if you had avoided entanglements. It is decidedly better to conduct your affairs upon strictly business principles, *i.e.*, let those for whom you work pay you in money, you in turn doing the same. In a word, you had better avoid everything that tends to efface your business rules.

Preserve a proper degree of gravity and dignity toward patients. Frivolous conduct, vulgar jokes, horse-play, clownish levity, unseasonable sportiveness,—

> "I love to laugh, though Care stands frowning by,
> And pale Misfortune rolls her meagre eye,"—

and bar-room familiarity are unprofessional, and tend to breed contempt and scandal. Discourage all attempts of roughs and toughs to rudely address you with a "Hallo, Doc!" or by your first name, or in any way to pass the limit of propriety with you. Show every one proper respect, and exact the same in return. Do not, however, understand me to advocate solemn pomposity, or to condemn good-natured pleasantry. Not so; for, when gentlemanly and in moderation, it is often very appropriate, and sometimes actually serves as a tonic to a patient's drooping spirits. If you, happily, possess a becoming earnestness of deportment, and at the same time wear a cheerful mien, it will be health to yourself and sunshine to your patients.

Avoid dining out with your patients and attending their tea or card parties. Eat as seldom as possible at their houses,— only when unavoidably detained there by cases of labor, convulsions, and the like. There is a tendency to conviviality and *abandon* around the festive board that has a leveling effect, and divests the physician of his legitimate prestige. It is far better to eat a cold repast at home than to occupy the best seat at the table and partake of the most savory viands of some patients. Let a physician once unbend himself among a certain class of people, and he risks a complete loss of their professional appreciation and regard.

When compelled by circumstances to accept a meal, if you

chance to be served alone, so much the better; if seated to eat with the family, be courteous, but somewhat reserved, and exhibit no uncalled-for levity, but simply endeavor to render yourself agreeable. Shun all *badinage* and gossip and undue extolling of the viands, and be careful to make no after allusions elsewhere to the "snowy cloth," the "delicious butter," the "juicy beefsteak," etc., as though you were a stranger to these.

Try to give satisfaction at your visits; show that you are anxious to relieve both the body and the mind of your patient, and you will not, can not, fail to succeed in your ambition to get practice. To do this fully you must, of course, feel and express a genuine interest in the case and in the effects of the remedies you are employing. Bear in mind, also, that, with any practitioner, the first essential to success is that he should command the confidence of his patients.

When necessary to scold or find fault with your patients or their attendants (as is often the case), either preface or follow what you say by explaining that you are *not scolding in anger*, but because you feel an earnest desire to have them do right for their own and others' sake. By thus prefacing your reproof you will completely disarm resentment, and, no matter how severe, all you say will be taken in good part.

If you are unmarried, it will, no doubt, be often cited against you; but the truth is, there is no great professional advantage gained by simply being married. The objection to most unmarried physicians is really not their celibacy, but their youthfulness, which may also be quoted against you even if married. It is true that the conversation and society of intelligent and virtuous females impart self-respect to man, and give elegance and tone to his manners; and for him to feel that the inspiring eye of such a one is upon him often inflames his soul with ambition to reach the highest goal and to win the greenest laurels. It is also true that "it is not good for man to be alone," and that every physician should, when his pecuniary circumstances justify the step, look out for a wise helpmate, settle down, and make for himself a home; but to marry with an eye to business only would be a very imprudent step, and entail expenses and responsibilities without corresponding benefits. Besides, you should keep both business and marriage on a higher plane.

"Without hearts there is but little home, and less happiness."

You will, in your professional career, often witness the misery, cares, and anxieties that flow from degrading the tender, half-human, half-divine bonds of marriage by entering into it simply to gratify lust, to obtain money, or from other ill-regulated passions, or from any other considerations whatever than pure love and congeniality of souls, and you had better seek no friend this side of heaven than risk the formation of the wrong kind of domestic relations yourself.

Everybody wants a lucky, conservative medical attendant,—

"Many funerals discredit a physician,"—

therefore, a series of dystocias, or of deaths in childbed, or of unsuccessful surgical operations, or of malignant cases, or cases of any kind that have terminated unsatisfactorily, often injuriously affect the physician for years by attaching to him—especially if he be a beginner—either charges of being blind to danger and to duty, or a *long-to-be-remembered* reputation for bad luck. If such a series unfortunately threatens you in the beginning of your practice, seek to strengthen yourself by consultations with able brethren.

No one can succeed fully without the favorable opinion of the gentle maids and acute matrons with whom he may be associated in the sick-room. They can be his best friends or his worst enemies. Women and children constitute four-fifths of all the population. Females have more sickness than males, and the females of every family are the autocrats of the sick-room, and have a potent voice in selecting the family physician. I have sometimes thought that the real secret why so many truly scientific physicians—to whom a patient is an object of scientific interest, just as a rock is to a geologist, or as a flower is to a botanist, who, more naturalists than physicians, love the rays of philosophy and the beams of science better than humanity; and with their eye at the end of the microscope, watch cases merely from a scientific point of view, or to study the action of medicine—very often decidedly lack popularity, and fail to get much practice, is that cold, unemotional, impassive logic and high theoretical attainments, however much admired abstractly, are not a certain guarantee of popular favor, since they are often attained at the expense of the endearing sentiments, and hence create none of those friendly ties upon which getting practice partly depends; but, on the contrary, are often associated with a deficiency of the qualities of head and heart which appeal to the weak side of woman—*her emotions*—and gain her favoring opinions, and secure her good will and word.

The power to impress those you meet with a favorable opinion of your adaptation to your calling is a potent and important factor. Discipline yourself by rigid self-examination whenever you have conducted yourself unsatisfactorily. This will teach you to conceal or eradicate your defects and faults, and to give prominence to your good qualities.

The faculty of being able to please, and thereby make friends of those who employ you in an emergency or tentatively, is likewise a power that you should carefully cultivate.

You will find, also, that remembrance of the names of children and of patients whom you see but rarely, and the ability to recall the salient points of former interviews with them, gives you a reputation for a good memory, and is a very useful adjunct to other qualities.

Three-fourths of all the population are children; and their likes and dislikes will control your destiny in many a family. Many people patronize various forms of quackery for no better reason than that "the children take it easily," knowing from experience that an attempt to give pills or bitter doses to refractory children who dislike compulsion, or spoilt children with resolute wills, whose nurses and mothers have taught them to look on "the doctor" as a barbarian or butcher, means a fight and a failure.

In your efforts to establish a practice you must not rely strongly on friendship or social influence, for men are influenced by self-interest, and your truest friends and acquaintances who knew you when you were a boy may prefer that you test your skill and gain your experience by attendance on others rather than on them or theirs.

Socially, you may be a great favorite while all are well; but when sickness occurs and death threatens, the principle of self-interest arises, and the impulses of friendship and kindred become dormant and do not determine the choice of of a physician. No member of any family circle will be spared, if any human power can save, and thoughtful persons, terrified at the possibility of losing the kind, provident husband, the beloved wife, blooming daughter, darling babe, dutiful son, or honored parent, as the case may be, instinctively send for the physician in whose skill they have most confidence. They go past Dr. Newstart, about whom they know too little; past Dr. Drinker, whose system requires so much stimulating, about whom they know too much; past Dr. Gay, Dr. Fickle. Dr. Aimless, Dr. Butterfly, Dr. Misfit, Dr. Strangeways, Dr. Blackleg, Dr. Phunnyman, Professor Halfsmart, and all others whose unprofessional demeanor proves them to be either unripe or unsuited to duties so delicate, so precious, so weighty as that of a family physician,—past all, till they reach Dr. Standbest, in whom their faith, their medical confidence, centres. Faith is the great controlling guide in choosing him who is to stand by what may be one's death-bed or the death-bed of one's loved ones. The greatest two elements of medical faith are: first, a belief on the part of the patient that you are anxious to do the best that can be done for him; second, that you are not only willing, but know how.

Be courteous and considerate to every one, especially when you are vexed or in a hurry; discourteous abruptness in physicians inflicts many useless wounds, some of which are difficult to heal. Politeness and courtesy are seed that cost nothing, can be planted anywhere, that always bear good fruit,—fruit that never withers. Resolve that you will cultivate them as long as you live.

When boys or young men come to you for assistance for their base-ball clubs, or their library, and the like, give something, and give it freely. If ladies ask you for a donation to aid the heathen (!!!) or to help buy a carpet for their church, for the relief of some one afflicted, or any other laudable object, give willingly and cheerfully. If the tiny boy or girl comes to sell a concert or other ticket, buy it laughingly; for contributions of this sort not only do good to others, but often prove to be a judicious professional investment for self. Were you to scowl and, with lengthened phiz, say "no!!" the young man and woman and the tiny boy would all unite in calling you "old stingy," and ever after avoid you.

There is a significant fact which you might not observe without having your attention called to it: it is that, after you get into full practice, your days, weeks, months, and years will flit by faster than those of other people, like the mists of the morning, because, as a physician, you will be incessantly engrossed with a medley of important absorbing cases, with the nature of your occupation constantly changing, and the flight of your time will consequently be almost magical.

You, yourself, are mortal; therefore, you should not only try to prolong your life, but to get as much out of it as you can, by seeking proper relaxations and amusements while the age for enjoying them remains:—

> "As we journey through life,
> Let us live by the way."

Many physicians, in the eager pursuit of business, foolishly postpone all relaxation from one time to another, intending to give up some of the hardest of their work and worst of their privations, and to fall back on their reputation for skill and experience, and then to take life easier, indulge in diversions, social amenities, and pleasure, when they get older,—in the autumn of life,—when the hair grows gray, etc., forgetting that

"An unlaid egg is an uncertain thing,"

and thus unwisely neglect to seek enjoyments till they lose all taste for them, till they know nothing, and are fit for nothing but

"Work! Work!! Work!!!"

and to wear out their lives in routine toil and drudgery daily and nightly, as the slave of the sick public, on the rough, hard, joyless treadmill of practice, hurrying up-stairs, down-stairs, from one sick-room to another, from some horrible sight to a stinking case, a death-bed, or a dangerous amputation, and from that to a repulsive obstetric case, or a puking baby, or some other kind of patient, weak, petulant, or exacting; often summoned unnecessarily, too, at unseasonable hours, or bored at home with office patients; then pouring over books and thinking day and night, till, from long-continued and extraordinary mental and physical exertions, they become prime candidates for one or the other of the physician's two afflictions,—organic heart disease or atheroma of the cerebral arteries,—then progressive heart-failure or apoplexy; next—death, as the penalty.

A little leisure, either to rest or to play, or rational amusement of any kind, soothes the troubled waters of professional life, and is a great blessing,—rec-reation is re-creation. Make it your rule to do as little work on Sundays and holidays as is consistent with your duty to others, and do not hold consultations on Sunday, except in cases of urgent necessity. The Sabbath, or day of rest, was instituted in Paradise, by God himself, and is a blessing to all. It is asserted that violating The Gospel of Rest, and working seven days in a week, instead of resting the tired brain, shattered nerves, and fatigued limbs on the seventh, shortens a life of three-score and ten by twenty years. I know a busy physician who, to protect himself, has a sign in his office saying "No Office Hours on Sunday." An occasional day's sport with rod or gun, or a summer trip, or an evening at a convivial meeting, or at the theatre, a change of occupation, or alternation of labor with ease of any kind, will work off nervousness and act as a refreshment to your labors; break the worries, frets, tumults, jarring, and cares of practice; vary the monotony of life, subdue mental tension, remove brain-weariness, soften the ups and downs, soothe mental excitement and nervous strain, conduce to health and longevity, and actually make you more philosophical and a better physician.

The cost of a pleasure trip or a few days' recreation is not, however, to be counted by the expense of your journey only, but you must also add the far heavier loss in practice, and the unmerited blame that is apt to follow being absent from those who need you.

Newspaper notices of your departure from the city for short sea-side, mountain, or other brief pleasure trips will, if allowed, have a disturbing and hurtful influence on your practice while you are away, and even after your return. Reporters are aware how such items injure physicians, and seldom publish them unless requested. The register-clerk of hotels where you register will, if asked to do so, omit announcing your arrival in the newspapers, which would publish your brief absence from business to the whole world.

If a professional friend is prevented from attending to his practice by sickness, or even by sickness or death in his family, it is just and proper to attend to practice for him without reward; but, if one goes in quest of pleasure and amusement, it is proper and just for him adequately to remunerate you or whomever else he gets to do his work.

When you get another to do your work, it will be much less laborious for him if you have your office patients sent from your office to his, instead of compelling him to spend or waste your stated office hours at your office; also, to have your family send each new call to him when received, instead of compelling him to call again and again to learn whether he is needed.

. After prolonged absence from home or recovery from long sickness, it is wise and perfectly ethical to announce the fact of your return to practice through the newspapers: "Dr. ———, No. — Street, has returned from his vacation (or recovered from his sickness), and will resume his practice immediately." Further than this keep your name out of the newspapers, and leave medical self-advertising to physicians who prefer to quack! quack! quack! and to rival pill-mongers.

Merchants and tradesmen attract customers by handbills and newspapers, and yet, even though they do exaggerate, such methods are not considered dishonest, because their customers are supposed to know something of the prices and quality of the articles offered; besides, they can go from dealer to dealer, to examine and compare before buying; but with quacking physicians the stranger has no such opportunity, no such safeguard; because their ads. and puffs tell only one-half of the story,—cures and successes,—and studiously omit the other half,—failures to cure and cases made worse,—and, since strangers allured to physicians by them can neither compare their skill, weigh their pretenses, nor gauge their honesty, all such resorts are deemed ethically wrong.

When you assume charge of a case for another physician, to look after during his sickness or absence from the city, or one of your own that has been under the care of a substitute while you were away, or that any one has attended in an emergency pending your arrival, take care to do as much good as possible for the patient, with as little harm as possible to the former attendant; continue his line of treatment, at least for awhile, if you can conscientiously do so. An abrupt, radical change, either in diagnosis, prognosis, or treatment, or designedly differing with him, either in opinion or practice, is both ungenerous and injurious to your co-worker. In such a case, if you believe something more should be given, instead of stopping his red or black medicine and ordering a white or yellow one, or his pills or capsules, and ordering tablets or powders, merely add yours to what is already being done, and thus avoid unpleasant reflections.

# CHAPTER IV.

*"The first step to wisdom is to be exempt from folly."*

ALWAYS entertain and show respect for your seniors in practice. There is probably no type of medical man more unworthy than coxcombical young Dr. Knowaheap, who overestimates himself, and considers that he is the most learned person in the profession; underestimates his seniors, and shows a corresponding contempt for them. Being fresh from college, and medicine being a progressive science, he may excel the older physicians in the use of the microscope, the stethoscope, and other severely scientific and technical points; but long experience has been their additional teacher, and they have a progressive clinical acquaintance with disease which gives an intuitive perception as to the choice of remedies, and in general makes them better logicians and much better practitioners; because there are peculiarities which belong to almost every disease, about which there is little or nothing to be learned from the books; and knowledge and skill derived from observation and experience far outweigh mere college learning, book knowledge, and specific formulas, to be learned by rote and applied by routine; and are more like part of one's very nature than that gotten from any other source, and are fixed indelibly on both one's senses and reason, to be brought forth again when needed. Remember, too, that although young physicians have recourse to scientific "extras," fine-drawn distinctions, and modern instrumental aids to diagnosis, and the very latest in treatment, more than do the older ones, yet in relying on these too much and on rational subjective symptoms and common methods, and especially on the unaided eye, too little, they are apt to forget the fact that the best part of every man's knowledge is that which he has acquired for himself, and that the art of curing disease owes more to sound judgment and common-sense bedside observation and experience than to anything else.

True, the practitioner who has grown gray and wrinkled in the profession is more apt to disregard the nicer pathological diagnosis, which defines the technical variety of the disease,—whether, for instance, a pneumonia is catarrhal, croupous, or interstitial,—and to be more attentive to the therapeutical diagnosis which indicates what the treatment should be; weighing the influence of age, season, rate of progress, complications, secondary affections, compensatory changes, and other clinical phenomena with a nicety that the junior with all his brains can never acquire from his text-books or in the lecture-room, and then, as with intuitive wisdom determining as to the best remedies for the mental and bodily sufferings of the patient before him,—reducing, evacuating, quieting, stimulating, or feeding him as foresight and experience have taught. The reputation of every physician is twofold,—one portion earned by himself, the other acquired from the general respectability and reputation of the profession, and public confidence in it; and such men—white-bearded or bald-headed or furrow-cheeked though they may be—have done very much to give our profession this honorable standing, and to smooth the way for others, and hence are certainly worthy of all respect.

On the other hand, the older physicians, having had their turn, and remembering the rough and difficult trials, the painful responsibilities, exhausting toils, and heart-rending doubts and anxieties and ill-treatments, and the blunders and sufferings and dearly-bought lessons of their own beginning, should show favor to their younger brothers without fear, and work side by side, with friendly feelings; for no matter how many aspirants appear, there is always enough work left for the older physician who has done his duty in the community; yea! the world is wide enough, and there is sickness and misery enough in it to keep every worthy hand and head and heart employed!

Life is a school for all. When you have been in practice

long enough to cultivate observation and to acquire an address in the management of the sick and to impress your patients with the fact that you have good common sense in everything and uncommonly good sense in medicine; have accurate judgment, and evolve practical wisdom out of your own brain; and that you know the duty of a physician in their sickness, and, in addition, are especially conversant with their moral and physical idiosyncrasies, such impressions will be of great advantage to you, and will make professional attention to them much easier.

"He knows the water best who has waded through it."

You will occasionally be employed in cases because you have long ago attended other members of the family in similar affections, and are very naturally supposed to know the peculiarities of their blood, and to understand the various points in the family constitution,—their temperament and hereditary tendencies and predispositions,—and to possess sovereign remedies for their relief.

You will find that the belief that you understand this or that person's constitution from brain to toe, from surface rind to innermost core, and know exactly what they require within and without, their likes and dislikes, is a powerful advantage,—one that gives you unusual prestige and a favorable chance to show your skill and to give them a still greater confidence.

Experience and skill are what the public especially seek in a physician; they are truly important, and everybody knows it. You should carefully try to show that you possess both. Of course, we all have aftersight, but far-seeing, prognostic foresight, and ability to correctly comprehend all the changes that have taken place between your medical visits, is what is needed. This is not described in your text-books, or furnished by lectures, but is sure to come from practical, dear-bought experience in diagnosing and combating diseases, and will develop and improve your judgment in every way, and enable you each year to see more fully into the very essence of your cases, and to foresee their events with increased clearness; and if you compel yourself to work faithfully and to develop the faculty of observation, every year will make you a better physician, and by the time you have worked and observed for ten or twelve years you will have attained the calmness of wisdom, and become clinically familiar with the symptoms and events of all the more common afflictions that confront us, and you will then know far better than now how to wave the Æsculapian wand, how to avoid former errors and mistakes, and also more easily and more exactly to shape your diagnosis, prognosis, and treatment in each.

In addition to the great advantage the older physicians have over the younger ones, from increased ability to discern the true nature and to foresee the probable degree and duration of grave and critical cases, and to give, concerning them, more discreet, definite, and truer opinions from the beginning, they can also, from experience, recognize and point out cases that are doubtful or likely to prove slow and tedious; they also have more staid judgment, and give better cautions and more solid precepts, and thereby relieve themselves of much anxiety and risk of blame. Such advantages naturally enhance their reputation, and enable them to reap the full value of their skill; give them a better address and greater confidence in themselves, and enable them to treat serious and tedious patients with steadiness, and, meanwhile, to retain confidence much better and much longer than younger physicians. This is the chief reason why physicians, sharpened by long practice, are less harassed in difficult cases by meddling officiousness from outsiders, and either dismissed or forced to call in a consultant, than younger ones, and why the practice of medicine becomes relatively easier and lighter every year. You will find that after you have practiced twelve or fifteen years; after many of the fine precepts, beautiful descriptions, and nice distinctions gotten from the professors at college have taken wings; after you have forgotten much of your theoretical Text-Book knowledge,—which was probably greater comparatively at graduation than it will ever be again,—your experience will give you an immense storehouse of practical facts that will be invaluable to you, and will often serve you in cases in which book-learning cannot; indeed, it is impossible to obtain from books alone sufficient knowledge of disease to make you a good practitioner. The possession of self-attained post-graduate knowledge gathered from the great book of Nature will make it appear to those around that you know what to do and how to do it, and is the kind that will make the public prefer you to your younger, less-experienced brother.

The public love to see a physician appear to understand his business fully and to discover the actual condition at a glance, or to know things intuitively; therefore, you must study and practice to be quick in diagnosis, and ever ready in the treatment of the common diseases and ordinary emergencies that will probably constitute nine-tenths of your practice.

Study carefully the laws of prognosis and probable duration of disease, for it is in these that young physicians are most deficient. Errors of prognosis are ordinarily far more damaging to the physician than errors of diagnosis and of treatment. Very few people can discover whether or not your diagnosis and treatment are correct, or otherwise judge the truth of your assertions or the justice of your reasoning; but if you say a patient will recover and he dies, or that he will die and he gets well, or that he will be sick a month and yet he gets up in three days, or that he will be well in three days and yet he is sick a month, everybody will see that you are wrong, and will very naturally infer that, as you were wrong in your prognosis, your diagnosis and treatment may have been equally so, and they will naturally seek some one else with more experience and keener foresight.

Skill in these things will enable you to foretell a favorable, a doubtful, or a fatal termination, and to foreknow the duration in a greatly increased proportion of your cases and save you a vast amount of anxiety.

In forming your prognosis use all five of your senses, if necessary; and be careful to ascertain not only the condition of the organ chiefly affected, but of the other vital organs also, since their condition and action may, in some degree, compensate for the lost or impaired functions of the diseased organ. Look also at the surroundings of your patient, and the nursing and attention he can command; and, lastly, learn to estimate, from the look, the voice, the groan, the cry, the breathing, the complexion, the gesture, and general aspect,—mental and physical,—his vital resistance to the disease (which differs in each individual), and then form your prognosis.

Bear this in mind: In your desire to soothe the fears of anxious relatives, do not wrong yourself in serious cases by pronouncing them lighter or less dangerous than they really are. Such mistakes often bring us sorrow and cause blame. The little pleurisy or the slight gastritis of to-day may be something greater to-morrow.

Never ask, as you enter to pay the first visit to a patient, the apparently simple, yet awkward, question, "What is the matter with you?" or salute him at any other visit with "How are you to-day?" or he will probably retort *that* is exactly what he wants you, the physician, to tell him.

Do not display the fact that you are a junior or a tyro working by reflected light, and thereby belittle yourself in the estimation of patients, by constantly quoting what this or that man's book says, or what such-and-such a medical celebrity thinks, or, worse still, by taking down your text-books before them, to learn what they say; as if you were deficient in readiness or in nerve, or had to rely on the opinions of others for all you know.

Also, never carry a ready-written prescription to a patient, as if copied from somebody else's book; rather commit it to memory, or jot it down in your visiting-list, to be glanced at and written off at the proper time.

The folly of blindly accepting or slavishly following the dicta of this or that master is nicely depicted by Molière in

*L'Amour Médecin,* Acte ii, Scène 2, where the following dialogue occurs between Dr. Tomès and Lisette:—

Tomès.—How is the coachman?

Lisette.—He is dead.

Tomès.—Dead?

Lisette.—Yes.

Tomès.—That is impossible.

Lisette.—It may be impossible, but it is so.

Tomès.—He cannot be dead, I say.

Lisette.—I tell you he is dead and buried.

Tomès.—You are mistaken.

Lisette.—I saw it.

Tomès.—It is impossible. Hippocrates says that such diseases do not terminate till the fourteenth or twenty-first day, and it is only six days since he was taken sick.

Lisette.—Hippocrates may say what he pleases, but the coachman is dead.

Take a lesson from this, and, if you have no experience of your own to guide you, adopt that of others; but remember that your patients of all shades, white and black, rich and poor, want to know what *you* think, and care but little for high-sounding names, or for what you have read in Hippocrates, Watson, Gross, or been told in lectures by your preceptor.

If you are determined to let people know you are inexperienced and have no opinion of your own, you should at least spare them the infliction of following you to the sources from, and through the processes by, which your borrowed opinions were obtained. If one is invited to dinner, he may imagine his host does not prepare it all himself, but he does not care to be taken down into the kitchen and through the pantries, and shown the pots and pans and rolling-pins, and to be introduced to the cooks and waiters, all to let him know exactly how the feast is prepared. One will feel much better entertained if he is, at the proper time, simply introduced to the table, smoking and groaning with its bounteous supply.

Remember this: Every one likes to believe that the physician is treating him by a regular plan rather than firing at random, more especially in diseases that are believed to depend on the blood or on any peculiar diathesis.

Make post-mortem examinations and scientific use of your opportunities, to confirm or correct your diagnosis and to become more familiar with the machinery of life, whenever fitting cases or questions as to the cause of death from unknown complaints present themselves; but never allow the inference that you are cutting or mangling the bodies of the dead to gratify idle curiosity; or to satisfy yourself alone, or to show that your feelings and emotions have passed through a process of hardening, or that it is a very great favor to be allowed to do it; but put it rather and emphatically on the higher ground that it is done for the benefit of science and in the interest of suffering humanity, and that it may be for the good of the very persons with whom you are then talking.

In making autopsies in private families never hurry, but take time and do them thoroughly, and be doubly careful to avoid unnecessary mutilation, and let your neatness and manner evince the greatest respect for their sleeping dead, and due regard to the feelings of those around, more especially if a promiscuous audience, or non-professional persons, are present, and, after concluding, hide all traces of your work as fully as possible, and then compare what you have discovered with your view of the case before death.

Bear in mind that all civilized and all savage nations respect the dead, and that the important uses of the dead to the living are the only, but all-sufficient, justification for human dissection.

Also, that it is morally and ethically wrong to consent to make a post-mortem examination of any one who has died under the care of a brother physician, at the solicitation of persons who, with mischief in their hearts, seek to disprove the diagnosis and disgrace the medical attendant; also, that when making autopsies, even in cases of accident or sudden death, the deceased person's regular medical attendant should, if possible, be invited to be present.

Out of respect to both the dead and the living, defer making post-mortem examinations for a few hours after death, if possible; as the hypostatic congestion that naturally follows death is often mistaken by the public for ante-mortem changes, and gives rise to the most wonderful stories of "a murder," "only in a trance," etc. It will always be well to point out to them its true nature and cause, and its utter lack of significance.

The useless and unjustifiable repetition of physiological and pathological experiments, made to illustrate already known facts, that require vivisection of animals is, by many, called cruel sport, and has received popular opprobrium, and will not add much to your reputation, if done with that in view, as such things are supposed to have been studied as far as needful in the laboratory and dissecting-room before leaving college. On the contrary, making clinical analyses of the urine and other fluids as an aid to diagnosis will not only lead to invaluable information regarding your patient's condition, but will be a great element in giving you popularity and professional respect.

Working with the microscope on proper occasions will not only increase your knowledge, but will also invest you, in the eyes of the public, with the benefits of a scientific reputation and its attendant advantages.

Obstetrical practice is undoubtedly, in some respects, desirable, especially in the beginning of professional life, as each case partakes somewhat of the nature of a battle in which the accoucheur is (thanks to Providence) nineteen times in twenty victorious, and his services are appreciated and extolled, and in future relied on, which gives him a retaining hold upon that patient, and paves the way to other permanent family practice. The inevitable and wearisome *waiting* at the bedside, however, entails a serious loss of time. Chance calls of any kind you can take or not at your discretion, but specific engagements, especially in obstetric cases, must be kept, day or night.

Should you ever get so overburdened with work that time is doubly precious with you, attending many obstetrical cases will so overtax your powers that it may become actually necessary, in self-defense, to restrict or withdraw from these and other time-consuming engagements, in order that you may get time to breathe and to attend to the rest of your patients with something like regularity, and to obtain your meals, sleep, etc., and do your writing and studying. Midwifery is a wearing and exhausting branch of medicine,—the hardest kind of hard work,—and in filth but little superior to the Nightman's; it seriously interferes with regular, healthy living, and is full of care and responsibility; and, although it does lead to other family practice, you will find, after some years, that the ordinary fees for attending cases of confinement are, on account of trouble, and anxiety while absent from them before or during labor, loss of time in waiting and consequent interference with the fulfillment of other duties and engagements, together with the nights of work, after days of toil, loss of sleep, risk of breaking down, etc., which they occasion, proportionately more meagre than in any other department of practice.

If you keep a daily record, you will probably find that nine-tenths of all your loss of rest is due to obstetrical cases.

When a woman engages you to attend her in confinement, write her name and address on one of your cards and hand it to her, with instructions to send it to you as soon as she feels that your services are likely to be required. This will emphasize the engagement, serve to remind her of the mutual obligation or contract, and make her more apt, when her time comes, to call you in than to call another physician, or to get a midwife with a view to save expense.

In spite of your having been engaged to attend a case, and being kept in suspense for weeks or months, you will sometimes learn that the confinement is over, that a midwife or granny was sent for, and the excuse will be that everything occurred in such a hurry that they could not wait for you, or had no messenger to send, or some other equally lame plea.

You will often be called upon in bad cases to do ugly work for midwives who have reached the limit of their obstetric knowledge, and for the sufferer's sake you should never refuse to go and assume charge of the case. Such occasions will afford you valuable opportunities to

"Do two hours' work in forty minutes,"

and to show the practical superiority of qualified physicians over the unskilled midwife and unpractised irregular, and also to enhance your position in the estimation of the public.

Pregnant women will sometimes want to make an Indian bargain with you beforehand, to come to them only in case their midwife fails. Of course, you should go to all cases where humanity calls, but you should hardly bargain with anybody beforehand to play second part to a midwife,—she to take the fee and *éclat*, if there is no trouble; you to take the care and responsibility for a nominal fee, if there is. You may be surprised to learn that it is now generally understood in many communities that every midwife has her regular medical referee to assist her in her complicated cases,—a one-sided bargain, which gives her the unearned *éclat*, if there is any, and him an undue proportion of worrying cases of tedious labor.

When you first visit a woman in labor it is a wise rule to ask her, among other questions, whether she has felt the motions of the child since labor began, that, in case she has not, and it is born dead, you may have some evidence that it was dead before you arrived, if such is truly the case. Also, if your examination of a primipara shows very unusual smallness of the vaginal orifice, incidentally mention the fact, and tell of the possibility of laceration of the perineum, that you may not be unjustly blamed if that should unavoidably occur.

In every primipara case of confinement, after delivering the child, be careful to call the mother's attention to the lump in her abdomen, and inform her that it is only the contracted womb. If you omit to do so, she may accidentally discover it, get greatly alarmed, and either await your visit with dread, or send for you post-haste.

The enlarged, pouched, or protuberant abdomen, that begins in many females a few months or years after confinement, believed by them to be enlargement of the womb caused by not being properly bandaged after labor, is due to adipose tissue accumulating in the omentum and in the abdominal walls, and also encumbering the abdominal viscera.

Attendance on patients at long distances has a tendency to derange and diminish your nearer practice, for while absent attending a remote call you may lose three nearer ones. Nor do distant visits, as a rule, pay in a pecuniary sense, but they do work an injustice to both physician and patient. Every one should have a family physician within reasonable calling distance. A few far-off patients will waste more time, break down more horseflesh, use up more carriages, harass you at unseasonable hours, keep you from bed, and expose you to bad weather more, and do more to make your life a hard one than all your other practice combined.

Keep your practice down to a number that you can properly attend; you can do this by sending in your bills promptly, weeding out worthless patients, circumscribing your field of practice, declining other than desirable obstetrical engagements, increasing your charges, etc. In refusing to take a case at a distance, or one that is likely to involve you as a witness in court contrary to your wish, or to accept an obstetric engagement, if you are really "*too busy*," assign that as your chief reason, as it is the least open to criticism and persuasibility of any that can be assigned.

Never offer as an excuse for neglect in visiting a patient, "I really forgot you;" to forget the sick is unpardonable.

Gonorrhœal and syphilitic cases are not very desirable on any account, except for the fees they bring; they are dirty, secret cases, and rather repel than attract their victims and their friends to the physician who attends them when they require a physician for other diseases. Accepting them will, however, often enable you to pick up a handsome cash office-fee.

Even when you are positive that a person has syphilis, it is not always judicious to say so.

*"All truth is not to be told at all times."*

Prudence will sometimes require you to reserve your opinion, but at the same time take care to give the proper treatment. Indeed, in practicing your profession, you will see and understand the results of many sinful habits and vicious courses to which you must appear more or less blind.

Be careful that your reputation for special interest in venereal diseases does not overshadow or eclipse other kinds, and give you the unenviable title of " P—x Doctor," and entail the social ostracism and loss of genteel family practice that would surely follow;—or that extra success in restoring the menses in females who suspect pregnancy does not bring an extra number of such cases to consult you, and give you the title of " Abortionist;"—or that attending an excessive proportion of sporting, courtezan, and bruiser classes does not give you the undesirable notoriety of having a "fancy practice;"—or, again, that perpetual and indiscriminate inquiring about the urine, and having it bottled for you (urology), does not earn for you the easily acquired title of " P—ss Doctor;" —

*" Four lovely berries moulded on one stem;"—*

or that a liver hobby, or a kidney hobby; or that the womb, or the stomach, does not become with you a scapegoat to be blamed for every obscure disease, give you the title of " crazyologist," and thus eventually impair your usefulness and harm your position as a practitioner.

You will find it much more pleasant and satisfactory to attend in some families than in others. From some you will constantly receive intelligent co-operation, and they will make charitable allowance for any little failure or shortcoming, while from others you will experience the reverse, and it will seem as if they want to perplex and harass you in every conceivable way, and as if they would make you feel, in attending them, that you were on trial for your life.

It is this harassment and continued feeling of personal involvement and perpetual anxiety, quite as much as overwork, that wears down the health and shortens the life of the physician. Bear this in mind, and let it be your philosophic rule and determination never unnecessarily to worry about anything you cannot help or avoid.

You cannot be too guarded in asking private questions, especially about diseases of an immodest nature, before people not in the confidence of the patient, unless they are clearly entitled to hear them; in such case, request all, and particularly those of the opposite sex, to leave the room before putting the questions. Be doubly cautious in this respect when your patient is a female, and the questions refer to marriage, menstruation, pregnancy, lactation, uterine affections, constipation, urinary derangements, or other delicate subjects, that her confidential secrets may not be exposed, or her modesty offended.

You will find it judicious also to avoid inquiring of a patient in stores or barber-shops, or on the street or other public places, in the presence of strangers, about his ailments, or those of patients at his home, unless they can be referred to without the slightest impropriety. Many persons are very sensitive in reference to their complaints and weaknesses, and captious concerning the time, place, and manner of asking about them.

So far as your influence can prevent, do not allow ill-judging and partial friends or patients to go about overpraising you, and speaking of you as a pet, etc. Inordinate praise, no matter from whom, is apt to arouse a corresponding dislike on the part of those who deem the praise either extravagant or misplaced; and such injudicious praise, while meaning well, will almost surely react against you, and do mischief. It might even arouse the angriest jealousy or hatred on the part of husbands, aunts, lovers, or others. Perfectly pure physicians have actually had to cease attending in families where such jealousy existed, to prevent causing domestic strife and estrangement.

It is also in very bad taste, and even injurious, for a wife or other near relative of a physician to praise him inordinately, and boast of his great skill and wonderful cases and cures; for people very naturally think such boasting is an attempt to send fish to his hook and grist to his mill. If done at all, it comes with more grace from comparative strangers.

Probably one of the most constantly useful faculties you could possess is the power of discovering, by a study of faces, etc., which are the *ruling spirits* in a family, and honestly securing their good-will, and keeping them satisfied with your services and your remedies. Also, learning the character and consequence of those who are likely to show dissatisfaction and give you trouble if opportunity offers.

While making your visits it is better, as a rule, to give your attention chiefly to the reports and conversation of the husband, if he be present, rather than to the wife, and to address your opinions, explanations, and remarks to him, or, in his absence, to whoever is at the head of those whom you meet in the sick-room, and to pay to all others only the respect that civility requires. If you do not do this sensitive "head ones" will feel ignored, and many will even get dissatisfied and create trouble for you.

Carefully avoid making communications to inquisitive or hostile nurses, or other prying mischief-makers, and, if necessary to answer their questions, do so in an ordinary voice, and not in seemingly confidential whispers.

When making a professional visit, banish all else from your mind but the case before you; and, no matter who may be present, let the patient, whether young or old, be the central object, and keep your thoughts and your conversation centred on him and his case. Both patients and their friends will naturally feel more anxious to know what you think of their cases, and to receive suggestions and advice for their benefit, than to hear anything else. If the conversation digress to other subjects, change it back to your patient and his case as soon as possible.

Adopt the same precautions, also, during consultations, and keep the conversation between you and your colleague on the case under consideration, and do not allow it to digress to religion, horses, politics, etc.; economy of time, both on your own and your patient's account, requires it, for, if a consultation lasts too long, it is apt either to terrify the patient and his friends, or induce a belief that you disagree, or are puzzled, or are talking horses or politics, either of which may undo you.

Skillfulness in changing or modifying your diagnosis or prognosis is all-important in all cases where a change has to be made.

In prognosticating the probable duration of a case, do not too hastily or definitely commit yourself; for whatever prognosis you foreshadow at the outset will, as a rule, be accepted. It is only when such prognosis is altered and assumes a graver form, or the duration which you have assigned for the case is much lengthened, that dissatisfaction arises. One of the greatest of all reproaches to medicine is that it is not an *exact* science; consequently, the practice of it must lack the element of certainty.

Do not get insulted at the foibles and infirmities and the hasty and angry words of your patients. Do not forget that the sick, unless their sensibilities are blunted by disease, are the most sensitive and the most selfish of mankind; and bear with the rude and discourteous treatment you will occasionally receive from the hysterical and the peevish, whose patience is down to zero and petulance up to a hundred; and from the frenzied, the eccentric, the unreasonable, the impulsive, the irritable, the weak, the excitable, and the low-spirited; and do not take anything a sick or silly person says in a paroxysm of anger, or during a period of despondency, or in great pain (or for want of sense) as a personal insult, unless you believe it is deliberately and willfully intended as such; in that case, do whatever self-respect seems to dictate.

Beware of confidants, and never become so partial to patients, or others, as to make them the repository of your professional or personal secrets. With our imperfect therapeutical means we cannot always attain perfect results or give complete satisfaction, and some of those whom you have served most faithfully, and regard as unlikely ever to change, will surprise and shock you by turning round and loudly decrying you. Bear the possibility of this ever in mind, and, while seeking to make your relations with your friends and patients cordial, frank, and free, take care to avoid telling secrets and making confessions that might be spitefully revealed, or put you in their power, if a rupture of friendship should ever occur.

When urgent necessity or great danger requires, you need not hesitate to do the most menial work for a suffering patient; but, unless these exist, pulling off your coat or collar, administering injections, giving baths, swaddling newborn babes, nursing the sick, dressing or undressing invalids, or moving about the room, rummaging drawers or ransacking cupboards in search of towels, muslin for bandages, spoons, goblets, etc., in the presence of other help, as a servant would, does not comport with professional dignity, and may be quoted as evidence that you are too familiar and lack proper self-respect. It is much better to ask for things you need and to let them be brought to you.

To be overassiduous in paying visits when no sufficient cause is apparent, or to be too deferential and superserviceable to those who think themselves extra good patients, is very injudicious, for as soon as one conceives himself to be your best patron, or that you are cultivating his good opinion too earnestly, or are calling oftener than he needs, he is almost sure to undervalue you and is apt to quit you.

A patient who is improving will generally be satisfied with a much shorter visit, a slighter examination, and less perfect attention in general than one who is not doing well, and especially if he is doing so well that, on taking your leave, you can express your emphatic satisfaction with his progress.

When a case is obscure, or in its initial stage, be cautious in expressing any positive or unguarded opinion; but in cases where you can safely do so, give free, honest expression to your diagnosis and prognosis. The habit of stating your views thoughtfully and candidly will compel you to search for the underlying cause and to analyze and weigh details closely; will also discipline your judgment and force you to study your cases, and to make a definite, clear, and discriminating diagnosis in any case coming before you, instead of rushing in and examining without thermometer, stethoscope, or other aids, lumping everything under the term "biliousness" (?) or "heavy cold," "heart trouble," "effects of malaria," "a complication of diseases," or some other ambiguous name, and prescribing, on the blunderbus principle,—that

"Mixed diseases must have mixed remedies,"—

after a moment's stare at the face, glance at the tongue, and touch of the wrist, and an equally hasty catechising of the patient concerning his most prominent symptoms, scarcely stopping to notice or study the minor ones, or waiting for a reply or maturing an opinion, and being off again in a moment,—as is too often done by careless routinists.

A careless or superficial examination of patients, or inattention to the history of tedious cases, neglecting to give a definite name to a disease, or calling it by a wrong one, or making light of it at the outset, has caused regret and loss of practice to many a physician.

It may be important for you to remember that, contrary to the popular belief, the art of medicine does not enable you, or any one else, to diagnose positively any of the eruptive fevers until their local manifestations appear. Where obscurity exists state it, for if you give a positive opinion and it turns out to be incorrect your reputation will suffer.

Frequently, when a case is grave, and you are being importuned to know whether you cannot do more, it is expedient casually to mention the things you deem contra-indicated,—

leeching, cupping, mustard, rubbing, baths, poultices, mopping the throat, electricity, etc.,—and tell why you have not had recourse to them, so that they may realize the fact that you are wide awake and have thought of them, but have good reasons for not using them.

Never pronounce any one's sickness feigned or trifling unless absolutely positive that it is so, and, moreover, never make fun of people sending for you, or being alarmed at what may to you appear trifling ailments or simple growths that will get well of themselves or that require no treatment, etc.; indeed, you should never joke, talk frivolously, or laugh about your patients or their sicknesses, either in their presence or elsewhere, nor taunt them about the trifling nature of their diseases. Some people will laugh off your treating their slight ailments slightingly, while secretly they will feel deeply hurt and resolve never to have you again, but see another physician, or perhaps resort to quack medicines for fear he might also laugh. Still another reason is that trifling ailments sometimes develop into serious diseases, and simple tumors may assume a malignant form, and their becoming so *through fatal loss of time* in recognizing the true nature of the disease is very apt to be ever after blamed on the jesting physician.

Again, never guarantee a cure, or certain success, or a sure recovery, for anything,—even a mosquito-bite; guarantee nothing, except that you know your duty and will do it; that if your patients will do their best you will do your best, and leave the result to God. Medicine is not a perfect science, nor is life a definite quantity. When pressed by persons who want the consolation of certainty to say that this, that, or the other case of sickness is not dangerous, it may in some instances be well to reply promptly, "Of course, no one can say there is no danger, *because* there is danger in everything, and any sickness, even a fly-bite or a pin-scratch, may prove fatal;" that thousands of lives have been lost when the danger did not seem greater, and tens of thousands have been rescued from danger that seemed more imminent, and that even a well person has no guarantee of life from one day to another. Also, remind the questioner that you, the physician, are but a mortal man, and that you do not possess supernatural wisdom, and do not hold in your hands the keys of life and death, and have not life-giving power; that your will and God's will may differ; and that, since medicine is not a life-insuring science, you cannot guarantee that any case of sickness may not develop some new symptom and become dangerous or have an unfavorable issue, or even end in death; then state what you think will be the probable issue of the case in question. Remember, too, that while every case presents a group of probabilities it is surrounded by a chain of possibilities,—a fact that you should not fail to mention,—and thus leave yourself a reasonable margin for uncertainties.

In giving death-certificates in mania-a-potû, syphilis, abortion, etc., never yield to importunities, or a false tenderness for family affliction, and substitute other pleasant-sounding terms that may possibly put you in a false position.

The laws everywhere confer on physicians honors, immunities, and judicial powers that are withheld from other classes. You are exempted from military and jury duty, and made an officer of the law over other citizens regarding insanity, vaccination, etc., and your certificates with reference to births, deaths, inability to attend court, serve on juries, do military duty, etc., are everywhere respected, and it is most certainly your civil and moral duty as a good citizen to comply cheerfully and promptly with all necessary restrictions and legal requirements,—to aid, rather than impede, enforcement of the laws.

Further, in giving certificates, it is best to certify, "In my opinion," etc. Indeed, not only is it more prudent, but far less pretentious, in expressing an opinion, written or oral, always to state simply, "I believe thus and so," or "In my opinion," etc. The fact that it is your opinion or belief no one can dispute, even though it should prove erroneous.

Be exceedingly cautious in giving certificates of insanity, with a view to consign patients to an insane asylum, and never yield to the importunity of mistaken or designing persons, and be guilty of the cruelty of depriving a fellow-creature of his rights, liberty, and property because he has temporary insanity, is harmlessly eccentric, or entertains some harmless crotchet, as, for instance, that he is a grandee, or that his legs are of glass, or that some lady of high rank is in love with him, while in all other respects he is sane and demeans himself and manages his property rationally. Be careful to distinguish between the really insane as contemplated by law, and those who are only seemingly so. Dissatisfied friends of such people sometimes give great trouble to accommodating physicians in these cases. Refuse to give certificates in all but clear cases, and keep a memorandum of all the facts in each.

Keep memoranda also, and be very guarded when called as a witness in will cases, suits for divorce, etc., with a view to protect yourself and, maybe, others against traitorous friends or designing enemies.

Rancorous feuds and venomous contests among those interested in wills and estates will occasionally arise and show you how selfish, and bitter, and mean, and unscrupulous mankind can be over dead people's old clothes and the dirty dollar. In all such disputes and family wars carefully avoid entanglement.

> "Gold begets in brethren hate,
> And gold doth friendships separate."

Never conceal the presence of a contagious disease from those around who are liable to contract it, or misrepresent small-pox as "measles," or cholera as "intestinal catarrh," or yellow fever as "the bilious," etc., as has been done, or you may very justly encounter the condemnation of the community at large. When your decisions concerning the presence of, or danger from, cases of infectious diseases (small-pox, scarlatina, typhus fever, etc.), and of their origin here, there, or in the other place, from local or domestic causes, militate against the wishes or supposed interests of hotel-keepers, store-keepers, boarding-house mistresses, etc., your views will, in all probability, be met by strong opposition. In such event, do not allow yourself to be brow-beaten into allowing any one to violate the laws relating to the public health. Your duty to the healthy is quite as great as to the diseased. Indeed, the protection of the public health is of far greater importance than the well-being of any individual. When, therefore, these or other dilemmas present themselves, adopt Davy Crockett's wise motto:—

"Be sure you are right, then go ahead.'

Be careful to prevent children, in whose family contagious disease exists, from infecting others by attending school, or otherwise mingling with those liable to contract it from them, and at the same time insist upon visitors being excluded. Take care, also, that its presence in hotels, stores, etc., is not kept secret at the public risk.

Never let people know that you are just from a case of small-pox, scarlet fever, measles, etc., or even that you are attending any contagious disease, for, in the event of any case occurring among those to whom you have made it known, blame will certainly be attached to you as having been the cause. If your practice is so full of such cases that you must tell it to somebody, tell the health authorities; indeed, the public good requires that you inform them anyhow.

After visiting contagious diseases, take care to disinfect your clothes by walking or riding in the open air; also, wash your hands with very hot water, or, if that be not at hand, hold them over the fire; disinfecting lotions, etc., may likewise be used with advantage; and further, if necessary, take a warm bath, or even a Turkish bath.

Oppose the conveyance of diphtheria, scarlet fever, measles, small-pox, cholera, yellow fever, typhus fever, and other contagious diseases in hacks, cars, and other public vehicles, and, if private ones are used, give specific instructions for their subsequent disinfection. Protest, also, against the attendance of friends at the funerals of persons who have died of such diseases, and insist on the imperative moral obligation that infecting particles from the dead must not be allowed to imperil the living. Make it your duty to enlighten the public on this melancholy subject!

Never keep a tongue-depressor for indiscriminate use, for, irrespective of the disgust that patients would naturally feel at having an instrument put into their mouths that had served a like purpose in many others, it might actually convey the virus of syphilis, diphtheria, etc., from one patient to another, and render you liable to very grave censure. When you wish to examine a throat at the patient's home, it is better to ask the nurse for a clean spoon, than to take a tongue-depressor from your pocket, or spatula from your case, and excite the patient's aversion and a lively curiosity among those present to know upon what kind of a case it was last used. At your office a clean, white, ivory paper-folder, kept lying on your desk, not only serves its usual purpose, but also answers very well for ordinary examinations; it is not at all disgusting, and is easily cleaned.

It is both right and proper that you should cheerfully lend a helping hand to aid a professional friend or great scientific superior, when he has any kind of case requiring it, and unite your friendly assistance with any member of the profession where humanity requires, or in any case that will either give you knowledge that you specially desire, or increase your reputation in a specialty in which you are interested; but be careful how you lend yourself as a jump-jack to bungling practitioners who would use you as a handicraftsman for their own benefit and repute, or run about assuming half (or all) of the responsibility in cases of fracture, wounds, luxations, etc., for third—or fourth—rate Sir Astley Coopers, or administer the chloroform, hold instruments, thread needles, use sponges, etc., for Dr. Rainbowe or Dr. Blockhead in their ambidextrous hacking and hewing, cutting and slashing exploits,—

> "In every act such mind the motto old:
> 'Be bold! Be bold! and everywhere, Be bold!'"—

unless you are properly compensated for the work you do and for the responsibility entailed.

Obsequiousness and subserviency should be heedfully avoided, for, not only will they never be of any solid benefit to your reputation, either in the eyes of the public or of the profession, but bad results in cases to which you are carried to assist the hero tend to depreciate you and to do you harm. It is much better for you, as an aspirant, to come out and stand before the world on your own foundation.

For somewhat similar reasons, if sickness, accident, or other providential contingency compel a neighboring physician to ask you to attend his patients for a short while, whatever you earn should be tendered to him; but if one incline to neglect his business for dissipation, or to pursue pleasure, he cannot expect you gratuitously to attend for him either long or often.

Preaching morals to dissolute patients seldom effects any practical good or makes the vicious virtuous, as moral distempers are usually too deeply rooted to be overcome by an appeal to the feelings; but you often can, by earnest, truthful advice and proper cautions, reconcile the estranged and calm the angry, and possibly turn the drunkard from his path. You can also exert the greatest influence upon patients who are being injured by indulging to excess in chewing, smoking, tippling, feasting, dancing, late hours, carousing, venery, the use of cosmetics, and other things that provoke or render them liable to disease. Your injunctions, indeed, in regard to these weaknesses, follies, and errors, if properly given, will be respectfully and attentively listened to, and will frequently be strictly obeyed.

When tipplers tell you that they intend to "swear off" for a definite period, advise them, instead of swearing off, to pledge their word neither to treat any one nor allow any one to treat them to liquor during the prescribed period. Such a pledge will be more manly and more apt to be observed.

"Eggs and oaths are easily broken."

Treating and the treater make seven drunkards in every ten.

The various quack bitters advertised and guaranteed to be "a wonderful discovery" are almost invariably some vile compound of bad rum or bad whisky, and are the origin of much drunkenness; you should point out the dangers and condemn their use. If a person *will* take alcoholic stimulants, advise him to take them "bare-footed"; then he will know what kind and how much he is taking.

A word about photographs may fitly close the chapter. If you adopt the habit of presenting your photograph to every one enamored of your professional skill, or of your manners, good looks, style of dress, etc., it will be the cause of many awkward dilemmas. Many patients who would swear by you one week will curse about you the next, perhaps charge that you have maltreated them, killed their children, crippled their wives, or done something else equally horrible. You will learn by melancholy experience that the minds of men (and of women too) are subject to rapid changes. Many who would regard your picture with highest esteem this month or this year would tear it down or give it to the hangman the next. Trifles light as air will sometimes serve to detach families from you; a whim, a word, a look, or a nod will sometimes break links that have been forming for years; indeed, even old patients whom you have served through cold and heat, darkness and light, rain and sunshine, will drop or dismiss you when they get ready, with less ceremony and less regret than you would an office-boy or an hostler.

## CHAPTER V.

"A mind fraught with integrity is the most august possession."

HAVE due and proper respect for religion. Your profession will frequently bring you into contact with the clergymen of various denominations. Do right, and you will not only find in them firm friends, but also your chief supporters in many of your most trying cases. The ministrations of a cheerful, discreet, and pious clergyman,—the Messenger and Servant of the Most High God,—

"With a face like a benediction,"—

who confines himself to his true vocation, *i.e.*, healing the maladies of the soul, applying salutary balms to the wounded conscience, binding up the broken-hearted, and comforting those who mourn, with remedies prescribed by the Great Physician, are sometimes more useful to a worn-out and irritated patient than medicine; and even in cases in which death is near and inevitable, calm resignation often takes the place of despondent fear and apprehension when the invalid is skillfully informed of the probability of death. In fact, when cheered and sustained by religion, and impressed with the belief that whatever may occur they are in God's hands, many of the careworn and sick show as little regret, upon being gently apprised of the probable near approach of death, as a traveler does when about to start on a pleasant journey.

"And hope, like the rainbow of summer,
Gives a promise of Lethe at last."

When summoned to attend cases of angina pectoris, aneurism, organic heart disease, desperate wounds, paralysis, or other serious injuries and diseases that create liability to sudden death, prudence may require you to conceal from the patient the danger of death, lest he at once lose all hope and be overcome by despair and tumultuous grief, which may exercise a grave and possibly fatal influence; these may even render mild diseases fatal in persons of a nervous constitution. In such cases, however, take care to give timely private warning to those especially interested, and never deceive them willfully, and never, so far as it is possible to avoid it, let any fellow-creature—high or low, rich or poor—pass away from life without your making the probability of such an event known to relatives, friends, or neighbors. Be also exceedingly careful in talking before children laid up with scarlatina, variola, rubeola, etc., of the danger of complications, or of their illness being serious or dangerous, and never terrify them, when they are to be operated upon, by loud preparations and an awful array of instruments; also, take care to banish from their minds the fear of hydrophobia, lock-jaw, etc., for many young children fully realize the meaning of death, and giving their cases such frightful importance would tend to terrify them. Be alike cautious in speaking within hearing of patients who seem to be sleeping, drunk, semi-comatose, etc. Bear in mind that a person, young or old, is not always asleep when his eyes are shut.

In this world of short meetings and long farewells it is just as natural to die as to be born, and every one's time must come sooner or later, and, although you can neither see what is written in the Book of Life nor detain the living soul when summoned by the Angel of Death, you will sometimes have cases in which you will seem to be vainly fighting death itself, and yet, to your astonishment, see the patient recover as if by resurrection; and, on the other hand, you will often discover that the patient has almost entered the gates of death, while friends around think that he is getting better, until your practical skill enables you to detect the gloomy fact. Be prepared, therefore, for such incidents.

In serious illness you can very properly prepare the way for the introduction of the clergyman, but you should never attempt to thrust your religious beliefs or disbeliefs or your political opinions upon patients who hold opposite views. Your specific duty is with the patient's body, and it is really no part of your duty to proselyte or to administer to the religious cravings of the sick. Every sect has a clergy of its own, to teach religion, to soothe the parting moments of the dying, and to make the sad event a useful lesson to the survivors, and to these you must leave the spiritual work. We are the physicians of the physical body, the temporal life; they are the physicians of the eternal life, of the soul. Do little or no theological talking or teaching, and, for the sake of spiritual decency, advance nothing that you do not believe and feel, that your lips may remain unaffected and your hands unspotted. Veil your views and confine your ministrations to the worldly welfare of patients, and never obtrude anything in religious matters that involves a creed antagonistic to that of the sufferer, and never belittle anything theological that the sick earnestly and honestly believe.

The momentous question of eternity is certainly more important than the transitory things that belong to earth; be, therefore, ever ready not only to allow, but, if need be, to advise patients to have spiritual comfort.

Religion does good not only hereafter, but here; indeed, the presence of religious faith pointing to a life of blessedness and immortality is a power that can assuage the keenest sorrow and suffering, and even make the avenues to death smooth; and if any physician does not recognize it, or if he feels contempt for religious practices or feelings, or speaks slightingly of religious beliefs, he (*need I add?*) lacks the A B C of philosophy and observation. Your own eyes will see many a poor, sick, woe-worn, despondent, and broken-hearted creature calmed in mind and soothed in body by its comforting and cheering influence, and aided by it to get well, if his ailments are at all curable; if not curable, his spiritual wants being supplied by the citation of the Bible promise that

> "These ruins shall be built again,
> And all this dust shall rise,"

and by the repetition of the Saviour's name. and the voice of prayer to a benign Father, he gets unfailing faith, patience, resignation, and hope from it, seems thankful to hear the worst, and becomes willing, or even anxious, for the hour of departure. Indeed, one who has strong religious hopes is more likely to survive a severe disease and go on to longevity than one who has not.

*"Hope is the pillar of the world."*

The involuntary, mechanical, automatic agitations, and seemingly anxious movements unconsciously made by many of the dying are popularly supposed to be attempts to communicate some remaining thoughts, or secrets, or special wish before death. In such cases do not fail to explain to the friends that a kind Providence has mercifully drawn the veil of unconsciousness around the dying, and that he is insensible to suffering; that the merchant has then forgotten his ships, the miser his gold, the millionaire his possessions, and the beggar his poverty.

The act of dying, in itself, is probably painless, and the last stages of painful diseases are, as a rule, less so than those that precede; but the dying struggle, though painless to the unconscious patient, is, nevertheless, often distressingly hurtful and harrowing to all who witness it.

Respect the religious belief of your patients; it is well that you, as a physician, whether a Roman Catholic or not, should be familiar with the following duties required at the hands of a physician by Catholic patients:—

When in attendance on Catholic families, be especially careful, in cases of dangerous illness, to warn the immediate friends of danger, that the sufferer may receive the last sacraments.

One of the seven sacraments of the Great Church of Rome is Extreme Unction. It is believed to purify the soul of the dying from any sin not previously expiated through other sacraments, and to give strength and grace for the death struggle.

This church teaches that moral responsibility begins at the

age of reason; therefore, Extreme Unction is necessary for all who have attained that age.

Extreme Unction is given but once in the same illness, but if the patient has recovered and shortly afterward has the same, or any other kind of dangerous sickness, this sacrament is again necessary.

Another of the seven sacraments of the Church of Rome, with which you should be familiar, is the Holy Eucharist.

The Holy Eucharist, sometimes called the Wafer, is believed to contain Christ's whole being,—his body, soul, and divinity. It may be administered frequently in all cases of sickness in which the patient is confined to the bed or to the house for any length of time, provided he has sufficient reason to make a full confession.

If the nature of your patient's disease is likely to render him unconscious, be careful to inform the family of the fact, so that the clergy may be sent for, and the confession be heard, and the Holy Eucharist given before the reasoning powers are obscured.

Those who are to receive the Holy Eucharist are required to fast, if possible, from midnight until they have received it; but if you consider that your patient's being without either food or medicine would be detrimental to his welfare the clergy should be informed.

Where there is excessive nausea and vomiting, the Holy Eucharist is either not given at all or given in the smallest quantity. To expose it to being vomited is a great irreverence.

Be also equally careful in Catholic families to administer, or have administered, conditional baptism to all children during or after birth, when there is the slightest reason to doubt their viability. The following are the conditions and details of conditional baptism: You, or any one else, whether a Roman Catholic or not, are allowed to administer it. A male adult is preferable to a female, and of course a Catholic, if one is at hand, to a non-Catholic. The baptism is administered as follows: After procuring a glass or cup of clean water (spring-water is designated, but hydrant-, or pump-, or any other kind of true and natural water will do), in a suitable manner say- "I baptize thee in the Name of the Father," precisely at the word "Father" *pouring* a small portion of the water upon the child's head; continue, "And of the Son," at the word "Son" *pouring* another small portion; again continue, "And of the Holy Ghost," and at the words "Holy Ghost" another small portion.

Bear in mind that in baptism every word must be uttered; were you to omit even an "of" the baptism would be insufficient. Also, remember that the water must be true and natural, and must be poured exactly whilst the formal words are pronounced. So very important are these details, that if you arrive after a midwife or other person has baptized the child, carefully ascertain whether she has observed the full form and used accurate language. If she has not, and the death seems impending, you should baptize it again. In such a case of doubt it is necessary to preface the formal words with, "If thou art not already baptized, I baptize," etc.

If in a midwifery case the child of Catholic parents is believed to be in danger of dying it must be baptized. If it is partly born, baptize on its head, if the head is presenting; if not, upon the hand, or foot, or any other part that is born. If no part is born, and you can reach the child through the vagina, the water must be applied to such part as can be touched. In all cases of unborn children, preface the regular form with the words, "If thou canst be baptized, I baptize," etc. In such a case apply the water to its body with a syringe, or by any other means by which the water will remain uncontaminated till it touches the child.

In Catholic families you will run great risk if you use the forceps before the child has been baptized; for if this be neglected, and the child be born dead, you will not readily be forgiven.

Remember that it is better that a Catholic patient should

be thrice prepared and not pass away than to go unprepared; therefore, if you err at all, let it be on the safe side.

*"A good judgment is a great thing."*

You should be especially careful to give timely warning of danger to all who have business of vital importance to transact, and to those who have the best right to know, in cases of sudden, serious illness; for instance, friends may have to be summoned, wills executed, and other arrangements made, and, more important still, the sick may also wish time to reflect on their awful situation and the more serious affairs of eternity.

In an adult with almost any sickness you can safely predict that a hearse will be at his door in a few days, at furthest, after the pulse has gradually increased (heart-failure) to 160. Also, if, after wounds or in acute sickness, he emaciates to two-fifths of his usual weight.

If you will observe closely, you will find that when a patient becomes fully and firmly impressed with the belief that he will die he is extremely apt to do so.

It is very much better for you to decline to leave your sphere as physician to become a witness to wills if called upon to do so, and especially in cases in which you are not fully satisfied of the mental capacity of the testator; and determinedly refuse to take part in, or in any way interfere with, the settlement or division of the estate of those whom you have professionally attended, as you may thereby incur the charge of misusing the opportunities afforded you by your position as a medical attendant. Of course, if a person whom you have served long, successfully, or faithfully, chooses to remember you for it in a corner of his will, if it is done without your connivance it will be lucky and all right.

In no case be a witness to or executor of a will when you are made a legatee or heir, as any legacy or pecuniary interest therein devised to you will be void.

When attending very serious cases, be careful to exhibit proper gravity and sincerity, and never try to excite hope in cases that are hopeless, or obtrude the cry of hope, hope! when you really see none. If, moreover, a very ill, sane adult really wishes to know his real situation: whether he is in great danger or likely to die, and pointedly asks you the question, answer him frankly and truthfully that he is a very sick man, and state fairly and fully the ground on which your opinion rests, and thus relieve yourself of the responsibility; but, if possible, couch your answer in kind and gentle language, so as not to appal and unadvisedly depress him by taking away all hope and substituting despair. At the same time, in expressing your opinion, give him all the hope, assurance, and sympathy you honestly can, and if you know anything favorable, either in his physical or spiritual condition, mention it as a solace. In anticipation of such painful incidents, it is well to be prepared with consolatory and advisory language adapted to the respective cases.

You are at liberty to be silent or to say but little regarding the nature or degree of a person's sickness, but, of course, let whatever you do say, whether much or little, be conscientiously truthful. You must not, can not put a falsehood in the place of the truth, not even when attending to the sick and dying; neither as a man nor as a physician must you, under any circumstances, sacrifice principle or honor for expediency, especially to a person in a terrible and trying situation. But you can, you must, as far as possible, soften the truth and blend it in a proper manner with feeling and sympathy.

You will find few who have the necessary fortitude and submissive resignation to seek to enjoy their remaining life after being told that their cases are incurable; you should be cautious, therefore, not suddenly to cut off all hope, even from those afflicted with lingering maladies,—tuberculosis, cancer, Bright's disease, and like cases,—in which death approaches slowly, like a creeping shadow, up to the last stage, knowing that persons with those diseases have plenty of time while sinking away—the face gradually grows thinner, their eyes dimmer, cheeks paler, pulse quicker, breath faster, limbs weaker, vitality and all else lower and on the wane—gradually to realize their true state. Indeed, you should not, in such cases, tell how many you have attended who died of the same disease, or give a merciless prognosis, containing, like a death-warrant, neither hope nor encouragement, unless you are prepared to be replaced by Dr. Bigsmoke, or Dr. Guller, or Prof. Oilytongue, who, unless by speaking more hopefully, can do no greater good than yourself. Be on your guard, therefore.

An imprudent and ill-timed remark may destroy life where buoyed-up hope would preserve it. To tell a patient that he has a grave-yard cough, or that he will never see spring again, or that you would not have his throat, or heart, or lungs, or liver, or kidneys for a thousand dollars, would add the depressing influence of despair to the sedative influence of the disease, and could not fail to destroy his moral will-power, and either murder him by inches or work serious injury to his case.

You may often prevent despondent and anxious patients—whose pulse or temperature has grown worse, or whose diseased lungs, heart, etc., you are examining—from asking you inconvenient questions that would necessitate a disclosure of your gloomy prognosis, by having ready on your tongue's end, questions regarding their appetite, sleep, state of the bowels, or something else to ask the moment you finish listening, counting, or testing; you thus give them no period of solemn silence, and so, temporarily at least, no chance for inopportune questioning as to what you find or think, or to see what your thoughts then show. Be prepared, however, for such at a later time.

We have to do not only with sickness, but the sick; not only with death, but the dying. It is, for several reasons, better never entirely to abandon a patient with consumption, cancer, etc., even though he be incurable, or in the last stages of a fatal malady; on the contrary, keep him on your list and visit him, at least occasionally, not only that you may give him all the comfort you can, by suggestions for the relief of pain and mental anguish, but that his sorrowing friends—present and absent—may have the very great consolation of knowing that their loved one will receive all necessary professional care and attention up to the very time the dark curtain falls.

In every stage of your career aim to convince the world that you, as a physician, are an apostle of hope,—

"White-handed hope, the hovering angel, gilt with golden wings,"—

of faith, of sympathy, of comfort, and of relief, and that your profession is not in league with the grim forces of death and mourning, but that, on the contrary, all its characteristics are indicative of health-giving and life-restoring power. Neither blue-eyed and rosy-cheeked Hygeia, nor her parent, Æsculapius, is represented as in tears, with the habiliments of mourning; but we see instead Æsculapius armed with serpents, the symbol of wisdom and convalescence, and Hygeia is seen affording to others warmth and succor,—a beautiful symbol of health and preventive medicine.

Remember that old Mr. Death is the physician's great antagonist, and that when he defeats your best efforts and extinguishes the spark of life your duty ends. Do not, then, essay (otherwise than mentally) to offer up a prayer, or make a prolonged stay to administer nervines to relatives or friends, or tender your services for promiscuous duties, such as carrying messages, going for the barber, or the undertaker, etc., but at the earliest fitting moment quietly withdraw.

Leave the laying out and the application of preservative fluids to the face and body of the deceased, and all such matters, to the undertaker and friends.

Abstain, also, from visiting the house of mourning to view the dead (unless from professional necessity or obligation), and, except when it is absolutely necessary, even avoid attendance at the funeral services of your deceased patients, or following their corpses to the grave.

More especially refrain from writing apologetic letters to

the bereaved, expressing self-reproach for failing to recognize this, that, or the other fact, or regret at not having followed a different course of treatment, and asking forgiveness. If there are any facts in connection with the case that call for an explanation, find occasion to communicate them verbally.

Ours is a chequered life, and we see humanity in all its varieties,—and are necessarily familiar with many of the most humiliating and revolting phases of human life,—the white and the black, the rich, the poor, the indigent, the putrid prostitute, the rascally outlaw, the swaggering rowdy (whose character can be read on his physiognomy), the depraved reprobate, and the sneaking thug; Swearing Joe, Thirsty Jack, Tough Moll, Joking Jim,—
"Puns and sarcasm he would pour forth at his own funeral,"—
Vulgar Sally, and Blackleg Tom (sin in satin and vice in velvet), will all be represented in the Babel of your practice. Attend anybody if you must, even in the haunts of idleness, shiftlessness, and drunkenness, as your mission is to all sick people,—the vicious as well as the virtuous; but, as far as possible, avoid disreputable places and the incurably wicked, and do not be hail-fellow-well-met with persons in whom the moral thermometer registers low, as they are more likely to prove a curse than a blessing; nevertheless, do not hesitate to do your duty to a suffering fellow-creature, however low in the scale of humanity and morality. At the same time remember that such Samsons and Delilahs respect no physician who does not fully respect himself, and take care to treat all such with ceremonious politeness.

Avoid all such deceptive tricks as to assure a timid patient that you will not lance his boil, but merely wish to examine it, and then suddenly doing what you assured him you would not attempt. Veracity should, in all the situations of life, and under all its circumstances, be your golden shield.

Endeavor to acquire and maintain a complete professional influence over all your patients, for unless you enjoy their confidence and respect you will have to contend not only against their physical condition, but their mental and moral also.

You have a perfect right to relinquish attendance on a case when you find your interest, or reputation, self-respect, limit of endurance, or other valid reason requires it; when you do so, give formal notice, and let the cause of your withdrawal be fully understood, that you may not be responsible for subsequent occurrences. It is better, however, to decline undesirable cases at the first interview, on the plea of great press of other work, than to take them, involve yourself, and then have to relinquish or neglect them.

Never refuse to rise from your bed to pay necessary night visits to patients; to do so would not only subject you to the poignant reflection that you had been recreant to the call of duty, but would also be unjust, in that it would put your duty on some other physician, and by delay cause the patient unnecessary suffering, possibly death; or it might even drive the messenger to a pharmacist for advice and medicine, or necessitate the calling in of an Irregular, or other undesirable person who could be caught up in the emergency. If, however, you will make it a rule to charge full *night-visit* fees for *all* visits made after bed-time, you will be spared much loss of rest and night exposure; calls to hurry two or three miles on cold, wintery nights, or before breakfast, on a half-run or dog-trot, because some one has sneezed, or to attend to other needless and harassing demands from frightened ailers, who fancy they are about to die, and from others who could have sent at a more seasonable time. *Unnecessary* night visits rob physicians of necessary rest, and even if they do bring extra fees, if your life is a busy one, they afford no fair equivalent for the risk to health from overwork and loss of sleep.

Be exceedingly cautious in accepting degraded or vicious patients, to be visited clandestinely, and in having married women or young females consult you secretly at your office,— especially if it be for vaginal or other private examinations,—

without knowledge of husband or parents. Also, be careful about attending patients suffering from the effects of intemperance or prostitution, under pretence that they have other than their real ailment, with the view to screen them, by misleading their friends or relatives.

Do not overvisit your patients, and be especially careful to pay but few visits to those with trifling injuries, uncomplicated cases of measles, mumps, whooping-cough, chicken-pox, etc. People observe and criticise a physician's course in all such cases, and, if he appear overattentive, they are apt to believe either that they are sicker than he admits, which will cause them great alarm, or that he is *nursing* and prolonging the case and running up a bill *unnecessarily*. It is sometimes an extremely delicate point to decide whether a patient needs another regular visit or not, and how soon; whether the case is of a kind in which changes are liable to be sudden and frequent or not. Practice should soon enable you to judge correctly. You must also learn the art of telling the proper time to cease attendance in different varieties of cases, so as to satisfy the patient and his friends that you are simply intent on discharging your duty.

Most people dread the expense of professional services, and excessive attention, numerous visits, and repeated changes of treatment are rarely appreciated; a physician who pays but few visits and yet cures is always popular. If you can acquire this habit, and gain the reputation of paying no unnecessary visits, it will be regarded as a special feature in your favor, and will almost double your practice. A good and the only proper rule is to visit your patient when, and only when, you conscientiously believe it to be necessary, whether once a day or once in seven days. Never go several times a day to observe the variation of symptoms or effect of treatment without pointing out the necessity for it.

Do not mix professional and social visits together; go either as a physician, and be one, or as a friend; and, above all else, avoid running in to visit patients unnecessarily because you "happen to be in the neighborhood." If you visit Tight-fist on such a pretext, and charge for it in the bill, you may be sharply criticised for making obtrusive visits and forcing un-asked civilities, and your bill possibly disputed. On the other hand, never visit a seriously ill patient so seldom, or so irregularly, as to lose sight of the details, or induce a belief that you are neglectful or indifferent.

Some well-to-do or overanxious people form an exception to this rule, and insist on your visiting them more frequently than is necessary, so as to almost live at their house during sickness, to observe progress, instruct attendants, etc., regardless of the additional expense; and, of course, you may gratify them, provided such attendance does not interfere with the fulfillment of your duty to other patients; but at the same time, if other than the patient himself will have to pay the bill, the person responsible should (if need be) be informed of the reason why the extra visits are made and of the unnecessary expense entailed. No blame can then attach to you.

During such frequent visits you should maintain a professional attitude, and avoid the habit of digressing from the patient to politics, the fashions, or other current topics; otherwise, he and his friends will be apt to lose confidence, after which the moral effect of your visits will be lost, you will be shorn of your influence, and may receive scant courtesy, and scarcely be honored or welcomed at your visits.

When visiting a patient, always let it be known whether and when you will visit him again; it will not only satisfy him, but prevent all uncertainty, and relieve the anxious expectancy of "The Doctor's Rap." Remember that to judge the condition or progress of some cases it is better to visit them at different periods of the day, or even at night, while others should be seen as nearly as possible at the same hour each day. When a case has so far convalesced as to make frequent visits unnecessary, and yet improves so slowly or irregularly as to make you fear an arrest of improvement or a relapse, it is better to keep an eye upon it by looking in occasionally, and letting it be known when you will call again, with an understanding that if in the meanwhile the patient becomes worse, or, on the other hand, if he gets so much better as to render your promised visit unnecessary, you shall be notified thereof. This plan is, for many reasons, better than quitting such cases more or less abruptly.

The old, chronic cases that beset our paths often do us great injury; for among the surest fruits of neglecting them will be the employment of quack medicines, or the entrance of an Irregular or charlatan, whom some busybody has pressed upon them during your absence. It is very mortifying to drop in to see a patient, after prolonged neglect, and see a big bottle of quack medicine, or a vial of pellets, or the two tasteless glasses sitting on the table beside him, and then to hear this or that story why they changed. When you first encounter a case already chronic, be frank and candid as to the time required, and as to doubts of possible cure, and use no disguise or equivocation, and make no rash promises.

To evince an earnestness and personal interest in your cases are potent, master qualities that inspire confidence and respect, and are often freely and readily accepted in place of superior skill. Seek therefore to imbue your mind with a feeling of genuine interest in your cases, and you cannot fail to show it in a thousand ways.

Make it a study to remember well all that is said or done at your visits, so that your line of conduct may be consistent throughout the case. Also, take care neither to betray a want of memory—

"Memory is the first faculty that age invades"—

or a lack of interest, for, were you to ask a patient, "What kind of medicine did I give you last?" or to hesitate in your questions, he and his friends would at once notice it, and suspect that you either felt but little interest in his case or suffered from a failing memory.

Study to make your address and manner such that **patients will not hesitate to open their hearts and fully impart to you their secrets and the nature, seat, and cause of their disease as fully as the pious Catholic would to his Father Confessor**. One of the greatest drawbacks to many physicians is that they fail to inspire complete confidence, and consequently patients neither intrust them with the secrets of their folly, simpleness, or wickedness, nor consult them in afflictions that create feelings of hesitancy or shame.

Have little or nothing to do with your patients' family squabbles or with their neighborhood quarrels, and do not let your wife or any one else know your professional secrets, or the private details of your cases, or of the methods or instruments used in their treatment, even though they be not secrets. Few persons like to have their foibles retailed around from house to house: what they said in their delirium, or how they shrank from leech-bites; how she "cut up" in her labor, or he gagged at a pill,—or to have other whims, fancies, or infirmities exposed.

"Whispers often separate chief friends."

Many persons labor under the impression that physicians who (injudiciously) allow their wives, for the benefit of fresh air, to ride around with them while making professional visits, relate to them all that has transpired during the visit after they drive away. Such, of course, is not the case; nevertheless, if people think so, the discomforting thought is the same whether it be true or not.

There is no end to the mortifications, compromises, and estrangements into which a physician's prying and babbling wife may lead him by her tittle-tattle. Nothing is more mortifying or vexatious to the feelings of sensitive patients than to hear that the details of their cases are being whispered about, as coming from the physician, his trumpet-tongued wife, or others whom he or she has told.

If you allow yourself to fall into the habit of giving out the latest inside news, or of speaking too freely ever of ordinary affections, or submit to be indiscriminately interviewed by Inquisitive Jack and Peeping Jenny, Mrs. Knowaheap, Mrs. Blabber, Mrs. Picklock, Longtongue, and other key-hole notables from Meddlesome Row, concerning your patients, your very silence in disreputable cases will betray them. The credit of whole families and the character of their individual members will sometimes be at stake, and unless you shut your eyes and close your mouth, lest you see and say too much, it may ruin them and involve you. Indeed, many persons would rather suffer or even die than be subjected to public shame or disgrace by an exposure of the affections they are laboring under; and some persons suffering with venereal diseases are so much afraid that their family physician might reveal the secret, that they would sooner allow the ravages of the disease, or apply to a quack, than run that risk of exposure and disgrace.

You will be, to a certain extent, an honorary member of every family you attend, and will be allowed to see people in a very different light from that in which other people see them, when their spirits are humbled in the hour of pain and the day of distress. The community, as a rule, view one another with a veil thrown over their moral and physical afflictions; their strong passions and feeble control; their blasted hopes and the sorrows that flow from their love and hatred; their poverty, their frailties, their crimes, their vexations, and their meanness; their cruel disappointments and rude mortifications, their follies and disasters, fears, delinquencies, and solicitudes. You will see the trappings of greatness and the cloak that hides deformity dropped, their infirmities and imperfections of mind and body with the veil uplifted, and the book of the heart wide open,— the homeless, the betrayed, the deserted, and the victims of intemperance, grief and joy, anger and shame, hope and despair.

"Mouth shut, eyes open."

You will hear conversations that it would be cruel to rehearse; you will become the repository of all kinds of moral and physical secrets. Keep them all, with Masonic fidelity.

Love, debt, guilt, shame, jealousy, grief, domestic trouble, superstition, poverty, anxiety, thirst for revenge, and the like may prey on the mind of a sick person, and actually convert a simple into an incurable malady. As such matters are apt to be concealed from you, it is necessary that you should bear in mind that they are important agents in the causation and intensification of disease, and be prepared for their early recognition.

Observe reticence at your visits, and do not allude to the private affairs of anybody from house to house. Let your lips be hermetically sealed to the fact that So-and-so has, or ever had, venereal disease, hæmorrhoids, fistula, ruptures, leucorrhœa, or constipation; or that abortions, private operations, etc., have taken place; or that any person has recourse to anodynes or stimulants; or that Mrs. Ohmy had a baby too soon after her marriage, or that Miss Awfulone or Miss Angelicus had one without being married at all; or that Mr. Badegg is addicted to secret immoralities, or that Moonlight or Sunrise has or has had a venereal disease; or that Allgood is not good at all, but has this, that, or the other bad habit. No matter how remote the time, if patients wish their secrets told let them be their own tale-bearer. You have no right to disclose the affairs of patients to any one without their consent.

But while judicious silence should be your general rule, it is your higher duty as a member of society, and to the laws, to expose and bring to justice abortionists,—

"Tremble, thou wretch, that hast within thee undivulged crimes, unwhipped of justice,"—

unprincipled quacks, and other heartless vampires, whether acting under cover of a diploma or not, whenever you meet with proof of their iniquitous work. But (even though morally certain thereof) never take a step for which you are not prepared to be held personally responsible, and never directly charge any one with dishonorable or criminal conduct on hearsay evidence, or unless you have at hand ample and unequivocal proof of his wrong-doing; for, if you are without it, the accused is sure to find a loop-hole of escape,—

"It is hard to catch a weasel asleep,"—

or to make an indignant denial, on the principle that

My "No" is as good as your "Yes,"

and cunningly bring against you a counter-charge of malicious persecution, with its legal consequences; after which he will *resume*—increased business—"*at the old stand.*"

In prescribing medicines for the sick, it is better to confine yourself to a limited number of remedies with the power and uses of which you are fully acquainted, than to employ a larger number of ill-understood ones; for which reason, you will act wisely in avoiding new remedies until their value as remedial agents has been satisfactorily proven. It may also be well to remember that the number of remedies actually required in combating disease is relatively small.

Memorize the rules for dosage, and keep in mind the maximum and minimum doses of every article you admit into your list of remedies.

Whenever you order unusually heavy doses of opiates, etc., instead of using the common signs, take care either to write the quantity out in full, or to underline both the name and quantity, or in some other unmistakable way show on the prescription that you are awake to all that is written. A good plan is to write at the bottom, "The above is just as intended." Again, when you write for a potent article that is but seldom used, it is well also to add its common name, that the pharmacist may feel no doubt as to what is intended. It is safer also to put the names of heavy-dosed patients on their prescriptions. When you order morphia, etc., in other than the ordinary doses, it will be well to have it made into pills and granules, and direct the pharmacist to "put them into a bottle." It is so unusual to dispense pills in a bottle, that it intimates to the compounder that the prescribed dose is not a blunder, but is as intended, and acts as a guard to patients and attendants against taking or giving them in mistake. When you prescribe pills, powders, etc., for sailors and other persons whose business renders them liable to get their medicines wet or wasted, it is better to direct them to be put into bottles or tin boxes instead of paper boxes.

You should have sufficient honesty and sufficient independence to do nothing, and to give nothing, to a patient when that is the proper course, but you may occasionally come across a patient with disordered imagination, who persists that he is bewitched, or that a pin or fish-bone is lodged in his throat, even after your careful examination has proven that there is none; or syphilophobic and anxious to take constitutional remedies after a chancroid; or morbidly afraid of hydrophobia or lockjaw; or the half-insane victim of this, that, or the other vagary or hallucination, and who cannot be convinced by all your assurances that his ailments or forebodings are imaginary. In such a case, when classical medicine and all else fails, as it is your object to cure, it may become not only justifiable, but as clearly your true and manly duty to employ such innocent remedies as are likely to relieve his heated and deluded imagination through psychological impressions, or the action of mind on mind, as it is to give certain medicines for a well-defined disease. Any remedy or expedient honestly intended to excite a definite expectation, or hope, in order to aid in relieving such a patient, is called a placebo. Giving a placebo in such a case is a very different thing from practicing upon and increasing a patient's fears, as is done by the charlatan. Despise not policy, but take care that your policy consists in honorable expedients for honest purposes.

A mental agent should, as a rule, be small and easy to take; the bromides, the valerianates, mild tonics, and other harmless remedies are sometimes given.

Should you ever have recourse to remedies intended to act chiefly through the mind, if you will take care to look your patient earnestly and steadily in the face, and give precise instructions concerning the time and mode of using them, they will do double good.

You will not only find that almost anything will relieve some of these mental cases, but will be further surprised to learn that it has evoked their enthusiasm, and that they are chanting its praise and vowing that they were cured of one or another awful thing by it. Some, indeed, who seem to be magically benefited by doses of—nothing—will actually credit them with saving their lives. What a sad comment on the boasted intelligence of the nineteenth century! What a pitiable fact for truth and science that flavored water, etc., often receive as much (presto, be gone!) praise as the soundest remedies! What a harvest such people supply for those who live by fleecing!

"Shame! Shame!! Shame!!!"

Of mental remedies make none but an honest and proper use (leave juggling and all that is dishonorable to disreputable pretenders), and, if you happen to have the appropriate remedy give it gratuitously and charge for advice only; if not, write a prescription for something that is not unusually expensive.

Never send a patient to a drug-store with a prescription for bread-pills or anything else you know to be inert. It is not right to cause any one to pay money for articles that have no intrinsic value; besides, if among all the simple tonics, nervines, etc., in the pharmacopœia you cannot select some recognized agent of more remedial value to a depressed patient than inerts, your resources must indeed be limited. Moreover, if a patient were to discover that he had not only been paying money for such inert, valueless articles as bread-pills or colored water, but wrongly exposed to the criticism and derision of the pharmacist, he could not help feeling victimized and indignant.

Let me here impress a caution: to believe too much and not to believe at all are both unfortunate mental conditions for those who practice medicine. Take care that your mind is not led into an exaggerated view of the importance and power of drugs. Bear in mind the example of the old woman of Paris who filled bottles with water from the river Seine, sold it as a cure-all, and heard of so many cures wrought by it on all sides that she died fully convinced by a throng of patients and a bushel of certificates of cure that the (polluted) water of that river was a sure cure for all the ills of the human race. Guard yourself, also, against the opposite and grave error that medicines are useless and unnecessary; for either view would materially impair, if not destroy, your fitness for the practical duties of your profession.

The very shape and fashion of medicine have changed with the present generation, and eight-ounce bottles and thirty-two ounce bowls of bad-tasting, purgative, expectorant, or diaphoretic medicines, wasting the strength in the beginning that is needed during convalescence, are (thank Providence!) seldom seen, for the vast majority of people are now sensible enough to avoid every-day medicine taking, and to use remedies only when sickness demands, and even then not too much; but taking a little "searching" medicine that "scours" four or five times, or a bottle of salts, or of cream of tartar, or ten and ten of calomel and jalap for "clearing the constitution," in the spring of the year, still has patrons, who believe in positive medication with positive results, just as the good housewife believes in spring and fall house-cleaning; and cathartics and other depleting remedies are still popular with the few who cling to the old FORTY-YEARS-AGO mania for purging, sweating, and cleaning the blood.

Such people always want to see and feel promptly and fully the action of medicines, and purge themselves entirely too often; and some of them think they could scarcely live a month unless they had almost turned themselves wrong side out with pills, salts, etc. Remember that when nature is relied upon the bowels ought to act daily, or at least freely once in two or three days; for when the bowels are naturally moved, the lower portion only of the intestinal canal is cleared out, and, during the interval before the next evacuation, the fæcal matter from above passes down and is in turn evacuated; but, when a purgative is taken, it sweeps out the entire alimentary canal, and of course such a scouring out is not required as often as the natural, though partial evacuation. For any adult who cannot have an evacuation without the aid of medicine, to give an aperient or purgative once in three or four days is sufficiently often.

Never tell patients too minutely how the prescribed medicine will act, as it may vary enough from your promise to disappoint them, and to brand you as a false prophet. There are also a few patients who would feel worried if you were plainly to tell them exactly what ails them.

Never solicit people, either by word or otherwise, to employ you; for such a course would tend to repel rather than attract them, and could not fail to deprive you of necessary respect and esteem. Besides, respect for yourself and the profession make it far better to wait until your professional acquaintance is sought.

Many people are naturally fickle and capricious, and cannot be depended on to adhere to you, even from one day to the next; no matter how earnestly one tries to serve and to satisfy them, they will quickly become wearied and disheartened, and will insist upon consultations even in the most trifling ailments; perhaps, also, change about with astonishing rapidity,—first from one physician to another, then maybe to a prescribing druggist or irregular practitioner, and will finally wind up with a quack or quack medicine. Others will adhere to you with steady confidence, through good and bad, with firm tenacity. You should, nevertheless, under any and all circumstances, base your hope of being retained and respected, no matter on what class of patients you are attending, upon the just and true foundation of deserving it. Do not, however, set your heart or faith on a continuance of the patronage and friendly influence of any one, for you will many a time be unceremoniously replaced, after days or weeks of unremitting attention, by those whom you know to be in head and heart far below you in everything that constitutes a good physician. Sometimes, after you have shown every attention, spent days of toil and sleepless nights, and done all that is possible, you will be unexpectedly and unjustly dropped by a family without reasonable courtesy or explanation, sometimes even when the patient is out of danger or nearly cured, and possibly be superseded by Dokter Lowebb, or Prof. Kornkutter, or little Dr. Bighead, or Docktur Killcow, or Dr. Bobtail (who spends half his time in trading horses and talking politics), or "an old woman," or an Irregular, who may at once change your diagnosis of "bilious remittent fever" to "malarial fever," or "typhoid," and change your sulph. quiniæ and mass hydrarg. to sulph. cinchoniæ and hydrarg. cum creta, and you have to submit to the humiliation, the icy ingratitude, and the wrong, without being in a position at all to resent it, and you will feel the force of Solomon's soliloquy: "If it befall me as it befalleth to the fools, why should I labor to be more wise?"

> "I have seen a stately cedar fall,
> And in its place a mushroom grow."

Poor people, when raised to wealth (from the dirt to delirium), often move from the obscure, old house, or dirty rooms, to a mansion in a different section, sell their old, shabby furniture, and buy new,—

> "But yesterday out of the egg, to-day they despise the shell,"—

pull off plain clothes and put on fine ones; and, as if to efface all the past, even the physician, who attended them in obscurity, and stood by them through everything, is also abandoned, and Prof. Highkite or Dr. Newmode is employed.

> "And thus the world goes round and round;
> Some go up, and some go down."

The faculty promptly to detect loss of confidence, or dissatisfaction with yourself or your remedies, is one of the acquirements which, if you do not already possess, you must seek to acquire. Bear in mind that continued suffering, protracted confinement, unsatisfied suspense, and disappointed expectation of convalescence all tend to produce impatience and dissatisfaction in the mind of the patient and his friends, and to create doubts of your knowledge, skill, or judgment,—for which due allowance should be made.

Even to proclaim a truth is not seldom attended with unpleasant consequences. Thus, you may deem it your duty to announce to a patient, or to some member of his family, that he has an incurable disease, or that he will positively die. Unless the fact is obvious to all, you will probably lose your patient, not by his death, but by his changing to some other physician, in the hope that he may reverse your verdict and give a more hopeful prognosis. Hence, before you give utterance to such opinion, make sure of the facts upon which it is based, for a hopeless prognosis deduced from insufficient evidence will inflict unnecessary pain on others and bring discredit on you. Nevertheless, being quite sure your judgment is correct, the possibility of the patient's quitting you should not deter you from giving timely intimation of the state of things, as it may be, for various reasons, of the highest importance that the patient and his family shall know it.

A patient has a legal right to dismiss you from a case at any time, but you can very justly expect that it shall not be done without cause, or without reasonable courtesy and explanation; and you have also a perfect right to relinquish attendance upon him at any time, provided it be done decently and in order. Indeed, you may sometimes find yourself so hampered, or harassed, or badly treated in a case, that either formally to retire from it or discontinue your visits are your only alternatives. When you discontinue your visits, give fair and timely notice to the person or persons chiefly interested, that you may not be liable for any bad consequences of neglect.

When you find it necessary to withdraw from a case, endeavor to do so in a courteous manner, for such withdrawal does not necessarily make it incumbent on you to break off all friendly relations with the family.

Whenever dismissed from a case, con over and carefully reflect upon the various circumstances that conspired to produce the dismissal, and the means by which you might have averted it, that, by self-analysis and self-training, you may acquire the art of doing your duty acceptably, and thus retain your patients.

Some people, indeed, who will almost idolize you as long as you are lucky and have neither unfortunate cases nor deaths in their families, will, as soon as either occurs, turn as rudely and maliciously against you, as if you kept the Book of Life and could control the hand of God.

When you are unjustifiably dismissed from a case, especially if it be to make room for an Irregular, or a foul quack, do not consent tamely to be cast aside in such a manner. Express your perfect willingness to retire, but, at the same time, make it known, in a courteous, gentlemanly manner, that you expected fair play and courteous treatment; that such dismissal grossly wounds your feelings, casts undeserved reflection on you, and injures your reputation in the eyes of the public, to none of which you can be indifferent. Such a protest will not only enable you to vent your mortification, disappointment, and disgust, but will also secure for you greatly increased respect, and, moreover, tend more effectually to counteract any injury likely to arise from your dismissal, than if you meekly submit without protest.

In acutely painful cases of tetanus, cholera morbus, etc., it may be found necessary to disregard the ordinary rules of dosage and give large, even heroic, doses of morphia, chloral, or other potent medicine, which must, moreover, be given promptly, as hours, or even minutes, may decide the result; care must, of course, be taken that the total quantity be within the limits of safety, and not sufficient to poison the patient. The following case will illustrate the point: A gentleman known to the author had a bad case of cholera morbus; a physician was called, who prescribed for him twelve opium pills, one to be taken every six hours. In that case the physician was fatally slow in his therapeutics, for long before the time to take the second pill had arrived the soul of that pain-racked sufferer had taken its flight to a land where medicine is not needed and six-hour intervals can do no harm. Take care to avoid his error, and never leave long intervals between the doses for patients suffering acute pain.

Bear in mind that an opiate that has power to relieve acute pain will do so within an hour; failure to do so necessitates a second or third dose. A dose of chloral will produce sleep within half an hour or so, if at all, and it is useless to wait longer before repeating it. When it is intended to keep a patient under the influence of opiates, it is necessary to repeat them every four hours or so, inasmuch as the effects of a dose begin to wear off after that time.

When opiates are no longer needed, the nausea that might follow their abrupt withdrawal may be prevented by continuing them in decreased doses at four-hour intervals, decreasing the dose each time to one-half of the preceding one.

There is a popular belief that opiates are given only to allay or relieve pain, not to cure the sickness. People should be made to understand that opiates are not only palliatives, but by controlling pain, lessening functional activity, etc., they are powerful curatives in a long list of diseases.

You will often recognize the character of a case, or see the patient's exact condition, before you ask a single question; yet the laity expect you to examine your patient at every visit. Let your first examination be careful and thorough; omit nothing that can shed light on the case, and never neglect the following five cardinal duties: to feel the pulse, to examine the tongue, and to inquire about the appetite, the sleep, and the bowels. No matter what the case may be, take care to attend to these and all other evident or special duties at every visit, else the patient may think—in your rush and hurry—he has not gotten the worth of his money, or is not properly attended.

Whenever symptoms render it probable that hernia, carcinoma uteri, Bright's disease, or heart disease is present; or that the throat is diphtheritic, or the ear occluded by wax; or that a tumor or an aneurism exists; or that one's femur is fractured within the capsular ligament, or his shoulder dislocated; or that a patient is pregnant, or has placenta prævia, or a uterine polypus; or fissure of the anus, or fistula; or that any other condition exists which, if overlooked, might cause unnecessary suffering or imperil the patient's life, and possibly consign you to the never-dying goadings of remorse for having overlooked a patent fact, or committed a grave mistake, or subject you to humiliation and disgrace, if discovered by another, who has come in,— all mind, all heart, all eye, all ear, all touch,—determined to show the wisdom of his being called, you should always make an immediate and thorough examination, and, if need be, gently hint to the patient or to the friends your suspicions and apprehensions.

"Too late, too late's the curse of life."

If you are careless or neglectful in these matters, you will often be surprised to see another, who has been called in consultation, or who has superseded you, discover the whole truth of the case; not so much from his superior skill, but because he made some examination or inquiry that you omitted. One of the most certain signs of a good and conscientious physician is to make an earnest, long-continued, and careful examination of the patient.

To mistake a tumor for pregnancy, or *vice versâ*, is one of the most mortifying and personally damaging errors of judgment that can well be made. To be attending a female who has been ailing for weeks and months, and who finally proves to be pregnant, is also very damaging, unless you have recognized and declared that fact; otherwise, her entire illness will be attributed to the pregnancy. In such cases some demur will probably be made to the payment of your fees.

Never ask an unnecessary question, yet be careful to make every inquiry essential to ascertain all the facts, and to satisfy the patient and others that you feel an interest in his case; if you neglect to do so, you will risk both an error and loss of confidence.

Prompt detection of dangerous changes, or of the approach of death, will not only shield you from blame, but will give you a certain kind of prestige if you point them out before the patient or his friends observe them.

Be careful never to speak of anything you may do for a patient as an experiment or from curiosity; for everybody is more or less opposed to physicians "trying experiments" upon themselves or theirs. For the same reason, it is unwise to give patients the sample bottle of new remedies sent to you for trial, or to let any one know that he is the *first* to whom you ever gave this or that medicine; or that his is the first case of the kind of fracture, or of small-pox, or of hernia, or of anything else you ever attended, or suspicion may take the place of confidence.

You should keep a register or a reference-book for collecting and retaining particularly good remedies, prescriptions for stubborn diseases, medical clippings, self-devised apparatus and expedients, self-discovered facts, and important things that you have seen, heard, read, or thought,—the substance of all. Such a record possesses continual interest and more value to its owner than any other book in his library; also a clinical case-book or a diary for recording the date, diagnosis, treatment, etc., of unusually important cases. Nothing impresses a patient suffering from a complicated or long-standing disease with a conviction that you feel an interest in him, and intend to try your utmost for him, so much as to know that you keep a daily or other regular record of his case. Besides, these records will become a store-house of facts, and furnish you important cases for relation at the societies or for publication in the journals.

When truth will allow, let your diagnosis either include the patient's belief or fully disprove it, that his mind may not distrust your opinion and treatment, and so tend to counteract your treatment.

You can more easily impress and permanently convince a doubting patient of a medical fact which militates against his wish or belief—for instance, that shortening is usual after fracture—by showing it to him in the books than by a hundred of your own verbal statements.

Demonstrations to a patient or his friends of certain diseases and injuries that admit of it, by comparing, in plain language, the affected parts to sound ones on a well-drawn pencil-sketch or diagram on a prescription paper, and patiently explaining simple facts that are not clear to them, gives great satisfaction, and makes them appreciate that you understand the case.

Study to be fertile in expedients, and be very, very, very slow to confess, or allow the inference, that you are hopelessly puzzled about a case, are at your wit's end, or have reached the limit of your resources.

Never be too sanguine of a patient's recovery from a serious affliction, and never give one up to die in acute disease unless the process of dissolution be actually in progress. Wiseacres say that "the only way to get well after a physician gives one up is to give him up" (?). Above all else, never withdraw from a case of acute or self-limiting disease because the patient is very ill, or seems as cold as ice, and more likely to die than to live; for the human system can often endure a great deal and still live; besides, it is always highly comforting to anxious relatives or friends to know that the physician, with his strong arm, kindly stands as a stay and support, ready and willing to do more, if the slightest opportunity occurs.

Icy coldness sometimes seems to make death inevitable within a few hours, and the physician hastens to announce it, when lo! reaction and high fever appear, whip up and revive the failing powers again, and make it seem as if he lost hope and abandoned the patient several days too soon.

If a patient be unable to swallow, think of the œsophageal tube; or if food taken into the stomach be not assimilated, continue your efforts with inunctions of codliver-oil, or oil and quinia;

also, by rectal alimentation, hypodermatic injections, etc., until he is either better or the breath is out of his body; for *nature*, by a crisis, or a vicarious function, or a compensatory process, or even the tardy action of the remedies you have already used, may turn the scale and let the life-power, or the special vitality of a vigorous constitution, rally and gain control over the disease at the very last hour. Under such circumstances, if you have hastily given up the case, and ingloriously abandoned the patient as hopeless, you will be justly mortified, while some brother-physician or an Irregular, or, may be, an old woman, who has stepped into the field at the lucky moment, will reap the glory of setting the laws of nature aside, and bringing back to life one already given up as dead.

# CHAPTER VI.

*"He is most free from danger who, even when safe, is on his guard."*

You will have to be on your guard against thousands of snags, pitfalls, and rocks, which lie hid in the paths and under currents to entrap and upset the unwary. When in doubt whether duty requires you to do a thing or not,—for instance, between doing nothing and a dangerous operation,—if all else be equal, remember that the sin of omission is, in appearance at least, not so great as the sin of commission.

A very safe guide, in determining what line to pursue in grave or puzzling cases, is to imagine yourself to be in the patient's dilemma, and then earnestly ask yourself, What would I have done?

We, of all men, need to be as wise as serpents and as harmless as doves. The most skillful physician may and often does get results that he deplores almost as much as the patient does, but which his sincerest desires and every care and his best judgment are powerless to prevent; therefore, in all ugly fractures, in capital operations, and all other serious cases, be they what they may, in which you think there is any danger of an unsatisfactory termination, and of your being blamed or sued in consequence for the result, never hesitate to seek professional aid. Having a second physician not only divides the great responsibility, but also constitutes each a witness of truth for the other, and, by making each the guardian of the other's character, will tend to divert hostile criticism, charges of unskillfulness, or of mis-diagnosis, and causeless suits for malpractice:—

*"Much caution does no harm."*

Keep in view, moreover, that the general community have the idea that physicians can and should restore broken bones and injured tissues, no matter what the injury may be, as perfectly as the Creator made them. Bear in mind, also, that when a fracture, or dislocation, or disfiguring wound, or accident of any kind, results in deformity, or shortening, or disablement,—contracted tendons, or rigid cicatrices, or a limp, or requires a cane or crutch,—there is danger of its being shown ever after as a botch or failure, and becoming a lasting and lingering libel on the reputation of the medical attendant. A badly set limb, or an unnecessary or bungling amputation, injures our whole profession, and the limb or stump may be held up in court in a suit for damages; therefore, the responsibility had better be divided. In this respect medical and surgical practice differ,—the results of sickness usually disappear, while those of unsuccessful or unfortunate surgery remain, and, if bad, may induce jealous rivals, tricky lawyers, or other conspirators to incite the patient to cast you into the fiery furnace of a lawsuit. Juries place a high estimate on the value of life and limb when sacrificed by supposed negligence or want of skill.

Among the reasons why many malpractice suits spring from surgical cases, and but few from medical ones, are these: Although one is as liable to a prosecution by the laws for medical as for surgical malpractice, it happens that medical cases are treated in a more private manner, and may each be treated in numerous ways,—some the very opposite of others,—and we are not compelled to give an exact name to every form and feature of disease, and a logical reason for every plan of treatment pursued; and persons interested, even though dissatisfied with the result, are not competent to judge of the physician's skill and treatment to the extent of a lawsuit; whereas, the amputation of limbs, the adjustment of fractures, the reduction of dislocations, the management of wounds, bandaging, etc., all depend on a knowledge of anatomy and on manipulative dexterity, and are all open to public observation and criticism; and the methods proper to pursue in any case are so well agreed upon by surgeons, and the results aimed at are so obvious, that even the vulgar may criticise and also prophesy. Having anatomy for his foundation and the science of surgery for his guide, the surgeon is expected to follow certain definite rules, to have infallible foresight, to overcome all surmountable difficulties, and to get a perfect result. If he does this, *éclat* awaits him, but if he gets a bad result, and is presumed to have omitted any duty, the painful task of vindicating himself against a lawsuit may follow.

The sooner your account with a dissatisfied patient is settled in one way or another, after your services are no longer required, the less likely you will be to have a lawsuit; and, if you do have one, the sooner after the services are rendered the better, while witnesses are still accessible and all the unfavorable influences are fully remembered.

Bear in mind that you have no right to do more to a patient under anæsthesia than it was agreed to do. To put a patient under chloroform to amputate a finger, or to remove an eye, and then to amputate the whole hand or both eyes, would give great reason for blame.

Keep your surgical knowledge at your fingers' ends; medical cases can be read up as they progress, but a broken limb, or a dislocated bone, or a wound, will not wait, and you must be ever ready to treat them correctly; and never forget that bandages, either too tightly applied or under which the tissues have swollen and constricted the circulation, have always been a fruitful source of blame and of lawsuits.

Always take the precaution, as well for your own as for the patient's protection, to examine carefully the action of the heart immediately before administering an anæsthetic; and to watch the respiration during the administration,—withdrawing the agent on the least approach of blueness of the face or lips. Take care to have another physician or a reliable assistant present in all cases in which it is necessary to produce anæsthesia, more especially if the patient be a female; and be ready to hang the patient head downward the moment weakening of the heart's action, or of respiration, or narcotism, requires. Also, have a third person present at all sexual examinations of females, especially if at your office, to disprove possible hallucinations regarding either improper language or actions, and to protect against scandal and the traps of designing people:—

"An enemy has sharp eyes and acute ears."

Here, I may say, beware of personal violence. Midnight desperadoes may, under pretence of sickness, decoy you into their traps, and then rob or murder you; or your brute, crazy with drink; or your homicidal maniac; or your fever-tossed patient who knows not what he does; or your lunatic with a delusion; or the infuriated fellow in whom you have made a wrong diagnosis or had a mistake in the medicine; or the unreasoning tiger in whose family you have had sad deaths, or an unfortunate case of surgery, or of unsatisfactory midwifery; or the insane wretch whom you have through kindness sent to an asylum; or the disappointed and desperate would-be suicide whom you have restored; the blackguard, the thug, the fanatic, the madman,—any blood-thirsty demon or member of the dangerous class may suddenly assault and try to maim or kill you.

You will not only have frequent lucky coincidences, which will give you unearned credit, but also occasional unlucky coincidences, in which the most unwelcome events will follow your therapeutics so closely as to seem to be due to them. Be ever ready to explain and defend.

All anæsthetics are dangerous; refuse to give them in trifling cases of minor surgery, or where a moment's fortitude on the part of the patient is all that is required. Such occasions do not justify the risk.

Never examine a female to ascertain whether she is illicitly pregnant, at the instance of parents or others, without her own full consent to the proceeding.

Such exclamations as "Oh no, Doctor! it cannot be that his arm (or leg) is broken, for he can work his fingers (or toes)" will often greet you when you pronounce that a bone is fractured. This error is due to the fact that people think that the fingers and toes are moved by the bones instead of the muscles. It sometimes becomes necessary to explain this in defense of the opinion you have expressed, or to silence some dissenter; every one can move his tongue, which has no bones.

Patients, especially those of the fair sex, whether virgin, wife, or widow, sometimes decline to allow the physician to make examinations that necessitate uncovering the body, or to allow him to see the underclothing or body, simply because they are unclean and unfit to be seen, while the physician erroneously supposes that the refusal is dictated by overshyness or modesty on account of the examination. In many such cases it is better, instead of insisting on an immediate examination, to respect their delicacy, or comfort, or convenience in the matter, and defer it to another time, to afford the desired opportunity for a change of linen, etc.

Occasionally, some suspicious father, or curious-minded husband, will show a determination to remain in the room during vaginal examinations, or during operations necessitating exposure of his wife's body, and you will feel tempted to ask him to retire, that he may escape the indelicate sight and *you* the embarrassment. If asked to retire, one or another might refuse to go, or do so with suspicious anger. The better plan in such cases is to inform the person that you are about to begin what your delicate duty requires you to do, and he will probably retire of his own accord, unless specially requested to remain. In the event of his antagonistic refusal to leave, it may become a question whether to proceed with the examination or operation, or to abandon or defer it.

Expertness in detecting and contravening the various kinds of scandal and calumny admits of cultivation to a great degree; so also does the ability to foresee and avoid entanglement with the captious, the mischief-maker, the silly tattler, the malicious liar, and the like.

Key-hole and back-window scandal-mongers, and lying

snakes in the grass, may also lie in ambush for you, trying to make something or much out of little or nothing :—

"Much broth is often made of little meat."

These must be met and checkmated by the most available means. To judge what is best to be done under the circumstances is at times a most annoying and puzzling question to be confronted.

Jealous midwives, chattering nurses, ignorant doctor-women, busy neighbors, and Job's comforters often exert a malign influence on patients, and tell tales and give instances of ignorance and of lack of attention, and circulate damaging falsehoods and rumors about physicians, that must be noticed and coped with according to the necessities of the case, but neither malice nor envy can do you any harm unless your own misdeeds prepare the way.

Tact and nice discernment in establishing and maintaining a proper attitude toward nurses and other attendants on the sick is a valuable power that will prevent or counteract many possible machinations. Love of approbation is natural; and to give attendants due credit on fitting occasions for the faithful discharge of their duty is not only just and gratifying to them, but tends to make them your firm friends. Such public indorsement, moreover, secures their further co-operation, and encourages them to do their best to maintain the reputation which you have given them.

A bad or ignorantly careless nurse may render a curable case fatal by improperly indulging the patient's appetite for food or drink, or by neglecting to give him medicine, drink, diet, etc., at the proper time, or in the manner directed; or by appropriating his food or robbing him of the stimulants directed to be given; or by subjecting him to excessive heat or cold, or giving him too much or too little fresh air; or by getting drunk, or becoming careless, etc.

The conciliation of anxious, captious, impatient, or dissatisfied friends of the sick, when the sickness is not progressing satisfactorily, requires no little skill and a comprehensive study of human nature.

In serious or rare cases, and particularly in such as engender great neighborhood or general excitement, if you indulge in confidential or semi-confidential whispers to the people or rabble, as the case may be; or incautiously give out daily bulletins to them regarding the patient's pulse, temperature, respiration, excretions, discharges, etc., it will often give rise to misrepresentation, or even to utter perversion of what you really did say or mean, and your statements may come back to you so changed as to necessitate tedious and irksome explanations from you. You will act wisely, therefore, in being ever on the alert to avoid danger. If it be necessary, express your opinion briefly to the proper persons, in writing, with the view to prevent its being misrepresented or perverted.

When a sick person puts himself under your care he gives you a responsible duty to perform, and he has no right to ask your advice without a sincere intention of following it; for if he then neglect or refuse to use your remedies, or obeys your instructions in a half-way or imperfect manner, he ties your hands and frustrates your efforts for his relief, and cannot hold you to full responsibility in the case. If, however, he will not or can not do exactly as you wish, and if no special danger exist, it is sometimes better, after drawing attention to the position in which you are placed (as a protection to yourself), to humor his antipathies, whims, or childish weaknesses, and modify or alter your therapeutics so as to meet his wishes and ability. This you can do in a good-natured way, without fully yielding to him or compromising your authority or dignity. The wishes, prejudices, impulses, and erroneous views of exceptional and fastidious patients must be studied and to a certain extent respected. To do this is a matter of policy, and is very different from yielding a question of principle. But if a patient be determined to use an improper or dangerous agent, you should of course refuse your sanction; or where you find it impossible to secure a faithful observance of your directions on the part of the nurse, or where they are bent on trying the latest nostrum, it may then become your duty to consider whether to go on with the case or retire:—

"Two captains sink the ship."

Never captiously oppose a remedy because it is suggested by a layman. An amateur nurse or the most ignorant person may make a wise suggestion; and laymen often talk excellent sense about medical facts which have come to their notice. Listen patiently to all sensible propositions, and, if simple and unobjectionable, you may find it judicious, if only for their moral effect, to utilize them in conjunction with your own particular treatment. Be frank in giving credit to any good idea, no matter by whom advanced; and when rejecting a remedy thus tendered, let it be known that your disapproval thereof arises from conviction and not from superciliousness. You may, also, in some cases humor a whim and sanction the use of harmless domestic (Grandmother) remedies,—herb-tea, mustard and other plasters, onions to the feet, etc., in conjunction with your more reliable agents.

Make it a rule to accord persons credit for well-meant deeds, even though they be valueless in themselves; also, when possible, to approve domestic treatment adopted before you were sent for; at least, do not condemn it in a violent or offensive manner. Listen patiently to those around while they relate how they did the best they knew, and do not pooh-pooh, shrug your shoulders, or smile sarcastically, and thus unfeelingly belittle their honest efforts to relieve the sufferer.

"Be to their faults a little blind,
And to their virtues very kind."

Your cordial approval of their simples, used in good faith with true and loving motives, will greatly redound to your credit, and greatly enhance your reputation for kindness and sympathy.

When attending certain classes of seriously ill patients,

*e.g.*, the wife of a druggist or the child of a physician, if there be any simple remedy in which they have great faith and which they wish to try, every consideration should incline you, unless there is some clear contra-indication, to freely acquiesce and allow it, in conjunction with your other means.

It will be a trying ordeal when, by accident, you meet an "old lady who has a never-failing salve," good for everything, from mosquito-bites up to tuberculosis. You will find her so full of faith in herself and in her great catholicon that neither reason nor ridicule can shake it. Be fair and reasonable with her, and treat her with courtesy and respect; but if you feign an attack of awe, or indiscreetly chop logic with her, and concede to her remedy any recognition beyond its actual merits, or meet her as an equal and take her into confidence or semi-partnership in the treatment of felons, ulcers, or wounds, you will make a mistake, and fill her matronly head as full of conceit, and of mischief, too, as the sea is of water.

Cultivate the quality of being a good listener, and let a patient tell his story in his own way, even though it be unnecessarily prolix or tedious. Hypochondriacs, who live in an endless midnight of gloom, the hysterical, the garrulous, the slightly insane, and various other kinds of incorrigible bores, who want you to "subscribe" for them, or who have become "manured" to suffering, and want to "insult" you on their cases, because they have heard you are a good "musician;" chronic wrecks, perpetual invalids, and troublesome "old women of both sexes" with a low level of health, looking ever on the sad side of life, will sometimes come to your office, and want to murder your time with annoying or unnecessary questions, or exaggerated descriptions of their ailments, for which a whole apothecary shop might be vainly prescribed; or to persecute you with the details of their business or their ancestry, or the history of their family affairs, with a whole Pandora's box of sighs and laments added, when you have no time to waste and yet are indisposed to be rude; and then tarry so long after the consultation is ended and the prescription given that you actually wish you could rise in your chair, make a polite bow, and open the exit door, else fly out at the window and escape from them.

Some of these you will have to freeze out by chilling coldness in their reception; or if you courteously let them know as they come in that time is very precious with you, they cannot deem you uncivil, and will be brief, unless they are unusually pachydermatous. If you are greatly annoyed by such visitors, have a placard posted with "Please be brief, as, being busy, I must divide my time." Another good plan is: at the first conclusion of a topic, or the first movement of the patient after the true business of his or her visit is finished, courteously to rise from your chair, as if you anticipated his or her rising to go.

To rid yourself of attending undesirable would-be patients will be one of the most difficult dilemmas that will confront you. If you are " *Too busy to attend,*" or " Not at home," these are probably the most unassailable of all reasons in such cases. To assume charge of a sick person and neglect him afterward is unjustifiable.

You have a right absolutely to decline to take charge of a case, but, if you do assume the duty, it constitutes a contract in which you agree to give proper attention and your best skill. To take charge and afterward neglect it is a great wrong. It is very much better at once to plead having too much other business, or any other true reason, and not take undesirable cases at all, than to take them, involve yourself, and afterward relinquish them, and expose yourself to criticism and abuse.

When you receive calls to cases that from any cause you can not or will not attend, you should at once notify them of that fact, that they may seek some other physician, so that the patient may be spared needless delay and you the annoyance of repeated messages and solicitations.

No one can blame you for not being at home when your services may chance to be needed, since you cannot be everywhere at once; but if you are at home, and quibble or refuse to respond to a call, you will sometimes be severely criticised, especially if the case should happen to go wrong in consequence of your not responding. It is much easier for a lawyer to refuse to take a client, or for a mechanic to decline a job or a merchant a sale, than for a physician to refuse to go to a case.

If you have a friend whom you would like to see called to a case that you decline, mention him to them by name. You can advise them to send for Dr. A., or B., or C., or D. If you have anything against Dr. E., be careful to avoid saying, "Do not send for Dr. E;" merely omit to mention him. Your silence will be condemnation sufficient. You are not bound to recommend a man, but you might have some subsequent unpleasantness were you to practically denounce him.

The chief objection to recommending persons whom you wish to cast off, to physicians whom you wish to aid, is that they are then quite sure still to hanker for you, and to involve you as a consultant or assistant to your *protégé* if things get serious; whereas, if, instead of recommending them to any particular person, you compel them to choose some one for themselves, you will get rid of them permanently.

You will occasionally encounter presumptuous patients, or their wiseacre friends or relatives,—

"All impudence and tongue,"—

whose ignorance is shown by the very fact that they do not know they are ignorant, who will, with a double meaning in every word, make meddling inquiries, examine and cross-question you, and rudely interject *their* opinions, or challenge you to controversy, or seek and presume to discuss your diagnosis and your remedies with you, thrust forward their own favorite doctors, or obtrude their prescriptions of food or medicines, and parley about the merits of various medicines. Such people are generally as full of doubts, beliefs, and theories as a lemon is of acid,—foreknowing and prejudicing all that you do, often destroying faith, and thwarting your every effort; may be, drawing the curtain aside after your back is turned, and exposing to everybody things that should rightly be regarded as your professional secrets. If you write a prescription for gonorrhœa, or cough, or, indeed, any other ailment, many a presumptuous patient or his keen friend will read it as fast as you have written it, and proceed to comment or argue on it.

"I ain't afeard to argify the topic with 'im."

You will often be harassed and cross-examined by such self-constituted Solomons, and compelled to resort to various expedients to satisfy or baffle them, and at the same time avoid collision with their whims, insinuations, and prejudices. In fact, from this cause, the good effects of mystery, hope, expectation, and will-power are of late almost entirely lost to regular physicians; all special confidence being sapped, all that you can expect in many cases is the gross physiological action of your medicines on the stomach and bowels of the patient, and prejudice and fear actually do much to thwart even that. Such meddling, ignorant gossips will make your duty difficult, and often actually aid in making curable diseases fatal.

When you prescribe a mixture of two or more articles that such a patient is familiar with, take care to make him understand that to judge the relative proportions needed of each ingredient is just as important as the ingredients themselves.

The presence of self-important sick-room critics, with jealous eyes and unbridled tongues, will, if you are at all timid, often impair or destroy your usefulness, by either diverting your mind from your case, or lessening your concentration upon it, and may even lead to mistakes in diagnosis or treatment. Consciousness of being watched by hostile nurse or visitor, who is hoping to detect some flaw, or a chance to make unfriendly criticism, or find fault merely to show that he is a zealous friend to the sick person, will, in many cases, embarrass your demeanor, and to some extent cloud your judgment, and, of course, mar your usefulness. More patients are visited to death by neighbors and friends than die by neglect.

It is better to leave your directions about medicine, food,

etc., with the nurse, or whoever may be in charge, than with the patient. Leave no room for unpleasant mistakes or queer blunders. Tell him in a concise, clear, distinct manner when and how every remedy is to be used, and how you expect it to act, and leave nothing to the discretion or guess-work of patients or nurses. "A few drops," "a little," "a pinch," "a sip," "a swallow," "a gulp," "a thimbleful," "about a mouthful," "a handful," "a cupful," "big as a peach," "the size of a bean," "every hour or two," etc., can each open the door for big mistakes. Be explicit as to whether the patient is to be aroused from sleep to take the medicine or not. Also, whether it is to be taken during the night at all, with or without water, etc.

Give all your directions at one time instead of in scattered fragments; take care to make them precise and complete, and if you have doubts whether they are fully understood, ask the person to whom you have given them to repeat them to you, or, if highly important, write them down.

Study so to control your countenance as to prevent your thoughts, embarrassments, and opinions from showing upon it during anxiety and emergencies, and be especially guarded in your manner, so that nervous and ill patients cannot detect in your flushing cheek, quivering eyelash, or faltering voice, unfavorable reflections about themselves which you wish to conceal; for while your eyes are fastened on them to appreciate their condition, theirs will be riveted on you to read their fate.

Make it a rule not to prescribe large quantities at a time in acute cases, as they may change from day to day, or even from hour to hour; it is far better to have the prescriptions repeated over and over again than to risk having half a bottle set aside untaken. It is always a nice point in practice to decide how large a quantity of medicine to order at a time. In many acute cases you may find it wiser to order only sufficient medicine to last from one visit to the next.

To set aside unused medicines and order others in such a way as not to impair confidence requires not a little clever management. In many cases where the effect of a remedy is exhausted and it is ceasing to be useful, or where any other indications for a change of treatment appear, it is better not to stop the old abruptly, as though it were wrong or doing harm, but, instead, to give instructions to discontinue it at —— o'clock and then begin with the new.

Patients will rarely complain of the cost of medicines that are taken, but they will observe the waste and criticise you when you set one half-used remedy aside and prescribe another. A good plan is to order the empty bottle in which one medicine was gotten to be washed and carried to get the next in. A medicine that has been discontinued is rarely again indicated. If, when you stop one remedy and order another, there be any prospect of its being used again later in the case, take care to mention it, as it will tend to avert the otherwise probable impression that there has been extravagance or misjudgment in prescribing.

Be also guarded against ordering patients to buy expensive instruments, reclining chairs, supporters, braces, atomizers, or other costly articles, unless you are very sure they will answer the purpose and will be used. It is anything but creditable to the physician to have people exhibiting this or that article that cost ——, ordered by him, but for one reason or another never used, and now referred to as a shameful instance of needless expense.

You will occasionally encounter patients who have been kept in a furnace of anxiety and terror for months or years (hell on earth) through the ignorance of some novice in the profession, who has examined their simple sore, or abrasion, and mistakenly pronounced them syphilitic, or through the deception of some rapacious and shameless quack, who, for the sake of fleecing, has falsely announced the existence of specific disease, to which their folly has unfortunately exposed them, and whose ravages and horrors their fears have painted to them in the blackest colors, when in fact they have really never had any true sign or symptom of that disease. It is torture enough for those who really have constitutional syphilis to go through life filled with gnawing remorse for the past, and consumed with horror for the future, without adding spurious cases. When examination proves that the case before you is not real syphilis, it is your highest duty to give such explanation and assurance as will fully banish the error from your patient's mind.

You will be sure to produce unnecessary alarm and distress, in the minds of those whose chests you examine, if, after going through your scientific movements, or, with your watch in one hand and the fingers of the other on their pulse, you tell them of "a slight deposit in the apex," "an abnormal resonance," "a *bruit de diable*," "râles," "a palpitation," "disordered rhythm," or other, to them, ominous symptom or harbinger. Be careful, therefore, to avoid saying or doing anything that will unnecessarily fix the mind of a patient on the character of his breathing, the action of his heart, etc.

You will occasionally meet with persons who were told years ago by Dr. Longface, or Dr. Ogre, or Dr. Sphynx, with a face as long as the bedstead, that their lungs were gone, and that they would not live a year; or that their measles would turn to consumption, or that they had the seeds of this or that affliction which would destroy life within such a time. Such unnecessary and unwise forebodings cast not a little discredit on the profession, and justify severe censure on the erring prophets who make them.

God only knows how many young women in our land are now nervous and dyspeptic, with hollow eyes, sunk in deep anxiety, tormented with apparitions of "womb complaint," which have no existence except in Dr. Spayemall's imagination or in Dr. Squintum's opinion; young women who—had not the subject been suggested to their minds—would have lived a life-time with scarcely a thought of such a thing as a womb.

The chief reason why womb-doctoring might tempt to dissimulation, and why there are so many spurious cases of womb disease, is obvious. When a man is told he has a luxated shoulder, or a cataract, or hernia, or cancer, he finds many ways by which to confirm or refute the physician's opinion, and he can also see what effect the treatment is having; but when a miserable, nervous woman, morbid on the mysterious subject of "womb disease," goes to Dr. Bugaboo or Dr. Fallopius, mounts his gynæcological table or chair, with its elevating mechanism, "gets examined," and is told, correctly or otherwise, with a solemn phiz, and with all the force of a proclamation, that her womb (like many noses) is "turned a little to one side," or "is down," "ulcerated," "dislocated," or "affected," or that she has pyosalpinx, or salpingitis, it tallies exactly with her fears; and, shrinking from both the expense and the exposure to be endured if she were to consult another physician, she naturally submits to the manipulations, long periods of treatment, and the monetary exactions of the physician or the charlatan who first made the examination,—possibly recovering from morbid states that never existed and paying for cures never performed.

"Small justice shown, and still less pity."

If there be a knave meaner than all others in the sight of God it must be the swindler who, void of moral sense, pretends to make operations on parts of the body the patient cannot see, or makes useless ones, or, with cold-hearted selfishness, exaggerates the nature of any case and terrifies the sufferer for dollars and cents.

"Remember him, the villain, righteous heaven !
In thy great day of vengeance."

It is also cruel to tell patients, without due explanation, that their trouble arises from the heart, or kidneys, or liver, or the lungs; or that they have the "liver complaint," or "kidney disease," or that their lungs are "affected," when there is only some slight or temporary affection of these organs. And it is still more cruel and unwise, and terrorizing, to predict immediate or impending death, even if you discover serious organic disease of the heart or the lungs. The duration of life will, in many cases, depend on circumstances that you cannot always foresee,—the cheerfulness or depression of the patient, his carefulness and prudence, the conservative powers of his system, the compensative efforts of nature, etc. You know that a man's liver, or his lungs, or his heart may be deranged this week and well the next; but many people think that if any of these organs are affected in any way it is necessarily permanent, and it gives them long and constant anxiety. Many people are at this moment living in as great anxiety as though a sword were suspended over them by a hair, because they were told long ago by old Dr. Vinegar, or Prof. Hasty, or Dr. Cowhelp, or Dr. Shallow that this or that organ was affected, without explanation being given of the functional or temporary character of the derangement. By explaining the difference between temporary ailments and those of a permanent character, or the difference between a functional and an organic affection, you will avoid magnifying real diseases or creating imaginary ones; infuse new joy into life, and insure many a patient perpetual sunshine in exchange for constant gloom. It is your duty, at least, to avoid all ambiguity of language in such cases.

Contrary to the belief of the laity, and of some physicians, sudden death rarely occurs in heart disease, except in aortic obstruction and regurgitation.

In nervous, hysterical, and impressible persons it is possible to convert a slight or even an imaginary complaint or functional trouble into a serious one by fixing their attention on the organ deranged; hence, in these cases, ignorance is bliss, and the physician should divert the mind of the patient as much as possible from the real or supposed seat of disease, even if he has to avoid letting him know exactly what he is being treated for.

Even our instruments of precision can be used in such a manner as to excite the brain, and become objects of study, fear, and dread. That excellent instrument, the (try life) clinical thermometer, often tells from day to day the unwelcome truth that fever continues, till it seems to the patient and those around that you do little else than measure how long he has to live, and they almost wish it had never been invented. Try, therefore, so to use it as to prevent, if possible, such ill results.

Many now rush to the vaginal speculum on the slightest pretext. Never make an examination with it unless a correct diagnosis imperatively demands.

Take especial care not to allow patients' attention to become unduly fixed on their urine. Some persons have a morbid tendency to watch this excretion, and only need a discouraging word from the physician to make them as anxious about their kidneys—with apprehensions of Bright's disease, diabetes, gravel, etc., arising before their distempered vision—as are some women about their wombs.

You will also have patients lacking in sound common sense—the salt of wisdom—(generally soft-bearded youths who have not yet cut their last molar teeth), who come tormented with evil forebodings over alleged conditions that are either imaginary or perfectly natural: some because they have discovered that their left testicle hangs lower than the right, or because their scrotum remains contracted or relaxed;—

"How green you are, and fresh, in this old world;"—

others terribly alarmed because they have, in examining themselves, discovered the little odoriferous glands on the posterior part of the glans penis, and imagine them to be chancres or cancers; others because the fear of disease, or of blackmail, or of a charge of bastardy, or self-accusation, or a reluctance to sin, or disgust for the (cow) female has thwarted their attempts to copulate with loose women and made them imagine themselves impotent. You will also occasionally be asked for advice by those about to marry, and by others newly married; and also by old sinners, who have burned the sexual candle too fast in youth and lived too fast in general, who find their passions or powers of flesh gone; and others who are almost crazy on account of this or that affliction, defect, or fear. In all such cases bear distinctly in mind that your opinions and advice are your capital, and do not fail to charge your *full fee*, even though you write no prescription. With such patients the charge is for banishing fears and anxieties and giving valuable information and satisfaction. No one has a right to tax your time and talent in any way without proper remuneration.

Be careful to warn all such people against the curse of falling into the hands of "Lost-manhood Quacks," Pseudo-scientific Anatomical Museum Impostors, with their terrifying plates and examples of venereal diseases, and other "friends of erring youth," who circulate pamphlets on the evils of spermatorrhœa, masturbation, etc., and do not fail to tell them of the mischief such skulking impostors inflict on their victim's health, and also of their merciless, never-ceasing voracity for money. A whole book could be filled in telling the various ways in which these human foxes wring money from their victims as you would water from a sponge.

> "Oh! for a whip in every honest hand,
> To lash such rascals naked through the land."

It is doubtful whether the various medical guide-books for the public—Dr. Quackem Wiseacre's "Family Medical Guide," Professor Scolasticus Lollypop's "Every One His Own Physician," and a host of others—ever do any one much good. It is certain, however, that people cannot understand them as they are intended to be understood, and that they do a great deal of harm, by filling men's minds with imaginary wisdom, and emboldening them to try their hands at doctoring cases that require a physician until either much suffering or permanent injury has been entailed, or possibly life itself sacrificed.

Are not such attempts to teach the laity how to treat diseases like trying to teach one how to read who does not know how to spell?

The eight or ten large papillæ seen upon the base of every one's tongue often occasion much anxiety, on being discovered by overanxious laymen, while looking into their throats for indications of syphilis, diphtheria, or ulcerations, and great relief is afforded when told that these are natural.

You will often be consulted by true syphilitics, who wish to know what would be the result of their marriage. To such never promise certain immunity against future outbreaks; and do not sanction marriage, unless three years at least have elapsed since they contracted syphilis, and at least two years since they had any indications of the disease. Even then they should marry only under hygienic and therapeutic restrictions.

When a patient, alarmed about his health, consults you, if you wish your opinion fully to satisfy him, *be earnest*, and let personal intentness to his case overshadow all that you say or do; take especial care not to divert his conversation from himself to extraneous subjects. If it be at your office, do not digress by showing him your paintings, or the toy steam-boat you are making, or by telling the latest bits of news or gossip, or the history of the good cigar or fine pipe you are smoking, or of the newspaper or novel you are reading, or of the cane you are twirling. If *he* divert the conversation from his case, bring him back to it at the first opportunity, and know nothing but your professional duty.

Never, under any circumstances, recommend sexual intercourse as a remedy for self-pollution, nocturnal emissions, hyperactivity of the sexual system, spermatorrhœa, hypochondriasis, disordered emotions, acne, unruly sexual excitement, priapism, prostatitis, or anything else. If those who are subject to these affections choose to run the risk of syphilis, or gonorrhœa, or bastardy, or exposure; or to commit rape, adultery, or self-pollution, or marry merely as a remedial agent for masturbation or nocturnal emissions, let it be on their own responsibility, not on yours. Perfect chastity is not only entirely compatible with good health; but I know of no disease, either of body or mind, in which sexual intercourse is essential as a remedy or palliative, and I do know that it is far better for all, male and female, to grow up pure in mind and body.

Bear in mind that night emissions recurring occasionally in young men partake of the nature of an overflow from seminal plethora, and are perfectly compatible with sound health. Young fellows, half-crazed with dread and remorse, will often consult you about these emissions, and you will find that almost every one attributes them to self-pollution in boyhood. The results of masturbation, however, are, as a rule, much less baneful and destructive than commonly supposed; when the nasty, unnatural habit is stopped, its results are, as a rule, quickly recovered from.

Consumptive females whose blood-making power is lessened by their disease naturally cease to menstruate. They then attribute their decline in health to the absence of the menses, whereas, the cessation is really due to the decline and consequent loss of blood-making power. When such patients appeal to you to restore their menses, you should explain why they have ceased, and why they will not return until their health and their blood-making power improve.

Consumptive persons sometimes have hectic fever so regularly at a certain hour, day after day, that they and their friends are persuaded that their sickness is malarial in character, and if you are not on the alert they may mislead you in your diagnosis, and into giving an erroneous opinion to that effect. If the administration of quinia has no specific effect on the periodic daily fever, in a weakly or health-broken person, you may suspect that it is hectic rather than malarial fever.

The popular belief that one is *looked* for consumption because a parent, or brother, or sister died of it, is true only in a limited sense. If the relative's disease were part of his law of development and were in his charter of life, so to speak, developing just when the rose-bud becomes the full-blown flower, it should indeed excite serious fears in every one who has the same charter, the same constitutional tendency. But if his disease began after his physical development was fully completed, or if it were brought on by an accidental cause, the law of heredity does not apply. One whose father, mother, sister, or brother died from phthisis, the sequence of bad hygiene, pneumonia, etc., is not thereby necessarily compromised, inasmuch as that variety is not hereditary unless his father had it at the time when he begot him, or his mother during her pregnancy or time of nursing. Only about thirty-six per cent. of those who die of consumption inherit it.

One person in every seven firmly believes that he has either heart disease or consumption, while those really affected with either are rarely willing to admit it, the consumptive generally to the last calling it a bad cold. You will find that the management of those who are actually suffering with one or the other is one of the most delicate questions in practice. When your opinion is invoked in these cases, do not examine or question them at all, unless you have time to do so thoroughly, for your primary opinion and treatment may influence their entire future course, and if anything be overlooked at the onset you may unwittingly induce a neglect of essential remedies until the patient is beyond their reach.

No wonder the mind dreads the great white plague, consumption, with its pale, ghastly white face, flushed yet pallid cheek, brilliant yet dimmed eye, husky cough, rattling lung and unsteady hand; for it is humanity's great destroyer. Many, once gay and happy, are to-day sunk into deep distress, because doomed by it to a certain and lingering death. It loves to scourge the young, the beautiful, the gentle, and the gifted,—

"Those made of Heaven's finest clay,"—

and this portion of every community is selected for its most intractable and rapid forms.

Valetudinarians, dreading every change of weather and every variation of temperature, almost invariably dress too warmly and heavily, and, in their anxiety to protect their bodies from cold, wear so much clothing that they shut all the air, sunlight, electric, and other health-giving influences from their bodies, overheat their skin, and keep it constantly relaxed, and, of course, reduce or destroy their natural resisting power, so that when the winds of heaven, or the cold air, or a draught, or even the gentle dews of night fall upon them, the result is like jumping from the climate of Cuba into that of Canada. No one, sick or well, should ever wear more clothes than are sufficient to keep him comfortable. Every ounce beyond that is unnecessary and enervating, and increases the cold-catching tendency.

People of the opposite extreme, knowing that cool bedrooms are healthy for hale and hearty persons, often carry children, subject to catarrhs and croup, and other invalids, from the warm rooms in which they have passed the day to cold sleeping-rooms, instead of giving them uniform warm air, day and night, till recovery takes place. It would even be less hurtful to reverse it, and keep them in a cold room while awake and in a warm one during sleep, because there is more nervous energy, and a person has greater resisting power while awake than during sleep, which makes the system more able to withstand cold. For instance, the butcher can attend at his exposed, fireless stall the coldest winter weather till midnight, and not even sneeze; but, were he to lie down on his stall and sleep during a similar period, he would probably get chilled and contract catarrhal pneumonia or rheumatism. It devolves on you to point out these and kindred dangers to patients who are risking them.

Register-heat, on account of its parching dryness, is bad for both sleeping- and sitting- rooms. You will often smile at seeing a small pan or cup of water simmering on a stove, or under a register that is pouring out a volume of overdry, impure air, while the inmates are blissfully believing that it is tempering and rendering pure and moist all the air passing over it. A very large wet towel or folded sheet hung over the opening, with its lower end in a basin of water, is much more effective.

Many newborn children are unwittingly exposed to the bad effects of cold from lack of knowledge on the part of those in charge. The popular belief is, that if the nurse puts plenty of clothes on a shivering babe she has done all that is needed;

whereas, if the little babe, whose heat-generating power is naturally very feeble, after a prolonged oiling, and soaping, and washing, and turning, and wiping, and powdering, and bandaging, be put into clothes in a cold condition, without further attention, hours or days may elapse before its feeble heat-making power can bring on a reaction, and warm its blue feet and cold nose. Ice is put into woollen cloths or blankets to prevent it from melting; cold bread wrapped in a blanket would never warm itself, but if warmed and then wrapped in a blanket it would retain the heat for some time. Take care, therefore, that the newborn babe is kept warm. As soon as it is dressed it should be nestled against its mother's bosom till warm; if this does not suffice, it should be kept near the fire till the requisite warmth has been imparted.

Remember that the act of nursing not only supplies the babe with nourishment, but also communicates a new supply of heat from the mother, and possibly electricity, or some other occult but useful influence; at any rate, it can do no harm to have all *hand-fed babes* nestled to some one's warm, bare breast at intervals of a few hours, in exact imitation of those that suckle.

The ancient custom of clothing infants in flannel, with woollen socks, during hot weather, creates discomfort and disposes to infantile sickness. Its harmfulness should, therefore, be made known to those whom you find following it.

There is a wide-spread popular error, participated in to some extent even by physicians, regarding the real object of lancing children's gums. When a physician lances or rubs a child's swollen gums, he does so not solely to let the tooth through, nor does he expect it instantly to pop through the opening, but his chief object is to sever the innumerable small nerves that ramify through the gum, and thus relieve the tension, pain, irritation, danger of convulsions, etc. No one should incise or rub through a child's gums, except when these evils are present, for so much prejudice exists on this subject among certain people, that if you lance their sick child's gums and in despite of it he dies, you will incur their malediction for doing it.

There is much less popular opposition to rubbing children's teeth through with a thimble, spoon-handle, or any other suitable article, than there is to lancing them; and the contused wound made by rubbing is less apt to reunite than a clean incision.

"Doctor, my child gets the phlegm up, but instead of spitting it out he swallows it again," is a stereotyped expression. If he does, it makes but little difference, inasmuch as he swallows it, not back into the windpipe or lungs, but into the stomach, where it becomes unimportant. It is, of course, unnatural for a child to spit before he is three years of age.

It is a popular belief that crossness in sick children is a favorable sign, and there often appears to be a great deal of truth in it, since it requires considerable strength and energy to exhibit crossness. The re-appearance of tears in the child's eyes when crying is also a favorable sign.

Never pooh! pooh! or make fun of mothers because they believe their children have worms, for in some instances they are correct in their opinions, and if you scout the idea and fail to give a trial remedy, you may be chagrined to learn that after leaving you they went to some drug-store, purchased a quack vermifuge, which, sure enough, brought away worms, and are exultingly telling it as proof that you were wrong and they were right. Such cases are a cause for blushing, and do one's reputation no good. It is better, when worms are suspected, to respect their opinions and desires, and give some harmless vermifuge, even though it do no other good than to test the fact and satisfy the mother. Mothers are often acute observers, even though they are not good prescribers.

It is better in all cases to allow a certain degree of weight to the opinions of the patient and his attendants, especially those who sit up with him at night; not that you should surrender your judgment to their exaggerated apprehensions or palpable errors, but, at least, listen calmly to what they say, and take in consideration their opinion in making up your own.

*"The nurse's tongue is privileged to talk."*

They see him the whole twenty-four hours, while you may see him but five or ten minutes a day. The apparently causeless fears and predictions of nurses and friends are sometimes surprisingly confirmed, and the self-sufficient physician's prophecies correspondingly unrealized.

Many persons think it is of no use to call a physician to sick infants, because they are unable to make any verbal communication, or to place their hands on the seat of their diseases to assist him in making a diagnosis. This opinion is erroneous, for the diseases of infants are usually simple in character, and their symptoms, being neither disguised, concealed, nor exaggerated, can be read and correctly treated by any wise physician.

Condemn the keeping of commodes in bedrooms, as they are a dangerous source of diphtheria, scarlatina, typhoid fever, and other filth diseases. Also, direct that the alvine discharges in all contagious diseases and the sputa of consumptives shall be either disinfected or destroyed.

Every worthy housewife courts the reputation of keeping her house clean, and one of the proofs of her skill is the absence from the premises of bed-bugs, fine-tooth-comb insects, roaches, and other vermin. Should you ever notice such things about a respectable patient's body, clothing, or bedroom, affect not to see them, for nothing is more deeply mortifying than to have anything of the kind noticed and pointed out by the family physician. Also, be oblivious to all mortifying accidents, immodest mistakes, and accidental exposures that may occur in the sick-chamber.

*"Doctors must often shut their eyes and close their ears."*

The terms scarlatina and scarlet rash are now in everybody's mouth, and are spoken of by the laity as harmless affections, under the belief that the word "scarlatina" means a slight affection, bearing about the same relation to scarlet fever that varioloid bears to variola. There is no such disease as scarlet rash, and the cases to which these terms are applied are usually either scarlet fever or rötheln (German measles), and, unless people are made to understand this, neglect of the necessary precautions and great damage may result.

Bringing out the eruption is one of nature's processes in measles, scarlatina, small-pox, etc., but there is no doubt that the large quantities of saffron-tea, ginger-toddy, hot lemonade, home-made wines, etc., used by grannies to bring them out, do more harm than good, by disordering the stomach, inflaming the eruption, etc. This "bringing out the eruption," when uncomplicated, had better be left somewhat to nature; when it is complicated, something more reliable than teas is indicated.

There is also a popular belief that all skin diseases result from humors in the blood that must be driven out, or, if already out, kept out, until killed by blood medicine, much the same as one would drive rats from their haunts and keep them out until annihilated. No patient will object to your driving his humor out, or killing it, but if he thinks you have simply driven it in, woe be to you should he subsequently have any severe or fatal sickness. In such cases it is well to give an internal remedy, whether local treatment is used or not. In some cases, where great fear or strong prejudice exhibits itself, it is even better to commence the internal treatment some days before beginning the local.

There is likewise a popular expectation of evil and a like readiness to blame the physician if any new symptom appear after he suddenly arrests or cures periodical bleedings, diarrhœas, foot-sweat, or chronic discharges of any kind.

Many persons suppose boils and various eruptions to be healthy. Not so; but, even if they were, most people will agree that some other mode of health is decidedly preferable. The belief is probably founded on the fact that during convalescence after certain serious diseases a crop of boils often appears, seemingly from a revival of the energies, or vital forces of the system, from the depressing influence of the disease. The fact of their appearance being coincident with reorganization and returning health probably accounts for the belief that boils and health naturally go together.

The high color of the urine occasioned by activity of the skin in patients whose sickness compels them to lie in warm beds or to keep in hot rooms, also seen in well people who perspire freely during warm weather, frequently creates alarm, and induces groundless fear that they have kidney disease. Explain to them how the functions of the skin and kidneys are in relation to each other, and that it makes but little difference whether the urine is scanty or abundant if it contain all the natural excreta and be simply deficient in water.

When a coin or other small foreign body is accidentally swallowed, some old lady is almost sure to recommend a dose of castor-oil or other purgative, thus liquefying the contents of the bowels and necessitating the passage of the extraneous body the entire length of the alimentary canal alone, instead of allowing the fæcal matter to remain as a mass to inclose it and prevent its corners and sharp points or edges from doing harm. When such an article is swallowed, do not interfere with the efforts of nature unless you feel sure she cannot expel it unaided.

When a person faints, those around run to assist him, and in those agitating moments instinctively raise his head, instead of lowering it as they should do, thus prolonging the syncope and endangering life.

In all cases where great debility and pallor are present, be careful to instruct the attendants to keep the patient's head low, and to prevent him from rising suddenly for any purpose, and from sitting up too long, or unwatched, for fear of fatal syncope.

"If the dog that bites a person goes mad, the one bitten will also," has caused many a valuable dog to be killed. The truth is, if the dog's mouth or teeth contain hydrophobic virus at the time of biting the person there is great risk of its being communicated; otherwise, there is no risk. If the dog be killed under the mere suspicion of having hydrophobia, all inquiries as to its madness are cut short, the disproof of the disease is rendered impossible, and the person bitten and his friends are left to all the terrors of uncertainty.

A hydrophobic dog is said never to live longer than ten days after it becomes so; therefore, if a dog bite a person, it should by all means be allowed to live beyond this period to ascertain beyond doubt whether it has hydrophobia, or is harmless (?).

Foolish persons will occasionally tell you, in a boastful manner, that they have no fear of contagious diseases, and will show either by word or manner that they entertain the belief that contagious diseases attack those who dread them and spare those who do not. It is well to teach such people that the laws of small-pox, syphilis, gonorrhœa, hydrophobia, typhus fever, and such affections, are very different from what they imagine; that fear cannot communicate them to timorous persons—man, woman, or child—who are not exposed to their influence, and that mere courage or absence of fear will not, can not, protect nurses or friends, old people or babes; nor braggarts, nor even persons who are unaware of their presence, if exposed to them.

You will often be asked what physicians carry or what they use to protect themselves against the epidemics and contagious diseases which they face. So it always should be, but so it sometimes is not, for we do not always escape; yet a fearless heart, a cheerful mind, and love of duty often do seem at such times to unite with hygienic precautions to protect us in our battles, while the cowardly fellow who flees from the danger would probably die if he were to stay.

"Epidemics kill hundreds, fear kills thousands."

There exists a popular prejudice against damp houses, leaky roofs, night air, dews, etc., which is probably carried entirely too far. Dampness is, of course, inimical to health when mold, absence of fresh air and sunlight, filth, noxious gases, or other defective conditions, or disease-producing elements are added to it; but neither life on board of vessels, nor the presence of excessive dampness, as in rainy weather, is in itself unhealthful. Dark houses and basement rooms, where the natural light does not reach, are, of course, unhealthy, and may breed microbes.

The low-spirited and the morbid will often refer to the fullness or emptiness of the veins on the back of their hands, and count the wrinkles, as proof that their blood is drying up, or that they need bleeding, or that they are consumptive. Explain to them the folly of such conclusions.

When liniments are applied to the extremities for swellings, pains, etc., the popular notion is, that to rub *outward*, toward the fingers and toes, is the proper and only way to carry off the disease, while physiology tells us to aid the impeded circulation by rubbing toward the heart,—a fact that it seems impossible to make people appreciate.

As purgatives after confinement, many physicians order simples: castor-oil, rhubarb, seidlitz-powders, etc., instead of writing regular prescriptions; it will be wise for you to follow the same rule and order castor-oil for a lying-in woman, or whatever other simple laxative she or her friends are accustomed to take, for if you give her a Latinized prescription for a purgative, and, as a coincidence, she has hyper-purgation, or puerperal fever, or hæmorrhages, or if syncope or anything else follow, she will be very apt to believe that your purgative was too strong, and caused her sickness; and if she happen to die you and it will be *blessed*.

The admission or exclusion of persons who wish to enter the sick-room, and allowing or forbidding conversation with ill patients, or within their hearing, require no little delicacy and tact. Exclude gossips and bores from the bedrooms of those who are seriously ill; also, kind but thoughtless idiots, who come to tell the sick how badly they look, or of the death or misfortunes of neighbors, or of those who had similar seizures; but try to manage the matter so as to engender no personal enmity.

Endeavor, also, to acquire expertness in answering anxious questions relative to such cases, but never attempt to exclude the parents, near relatives, religious advisers, or other privileged persons from the room of any one who is seriously ill, and never interdict Bibles, prayers, and religious exercises, except for the most urgent and obvious reasons, such as, when he is highly delirious, or has just taken an urgently needed sleeping potion, or where he is determinedly opposed to the introduction of religious conversation; else you may raise among them a very natural whirlwind of indignation against you.

To interrupt public business and travel by roping or barricading the streets, because somebody is sick in one of the houses, is seldom either necessary or justifiable, as most dwellings have back rooms into which the sick can be taken out of reach of the noise of travel; make it a point to advise the latter course instead of the former. Where the former is not feasible, a good bed of spent tan on the street will completely prevent the rattle of passing vehicles, and show everybody that a person is sick whom noise will injure.

Never ask the age of a patient more than once during attendance on his case. Take care, also, neither to ask any question twice at the same visit, nor to do anything else that would indicate abstraction, lack of memory, or incompetence.

You will find that patients will be inspired with more faith in a prescription if you begin to write it with an air of decision immediately after receiving, to a more or less important question, an answer which your manner indicates is what you expected.

Take care to ask all necessary questions before you commence to write your prescription, lest the patient think that, in forming your opinion, you have not included the additional information or taken it in consideration in writing the prescription, or that your treatment was determined on before you obtained all the facts. Make it a rule, also, to ask no further questions after prescribing. It is well to terminate a visit as soon after prescribing as can be properly done.

# CHAPTER VII.

*"The successful man is the man who knows human nature, as well as his profession."*

EVERY minute spent in studying how to make your remedies agreeable will be more advantageous to you than half an hour of any other kind of study. Whoever, in these days, prescribes nauseous, repellant medicine in ordinary cases injures both himself and his profession, and is deficient in one of the most simple and essential requirements. Indeed, one of the great drawbacks to young physicians, and one of the chief reasons why they fail to render more practical assistance to their older brethren in superseding pleasant quackery, is that, having their attention riveted on their cases and studying more about getting them *safely* than *comfortably* through their ailments, and being anxious to get the specific physiological effects of medicines quickly and fully, they too often give them in crude forms, forgetting that the majority of sick people are fastidious, and have likes and dislikes that must be respected.

A great and almost universal mistake that *regular* physicians make is in supposing that when people send for them it is for the sole purpose of having medicine prescribed. Many people, not being judges as to what cases need medicines and what do not, are much more anxious to see one who does know, have a talk with him, and get an opinion of the nature, danger, probable course, and result of their cases, and words of assurance from him, and some simple remedy, if necessary, than to be drenched at every pore, or begin a medicine-taking siege, or a bombardment with gross drugs.

Make special endeavors to retain every medicine-hater who chances to fall into your hands. Such *incorrigibles* had better be under your care, with rational supervision and small doses of good treatment, than to be paying some one else for harmful quackery or fantastic nonsense.

Keep ever in your mind that mankind has both a material and a spiritual nature, and that different people seem to be made of different and almost opposite qualities; some seem to be two-thirds spiritual and one-third animal, others seem to be but one-third spiritual and two-thirds animal, between which are all intermediate grades. If you treat all these alike, you will certainly fail. The mind belongs to the legitimate domain of the physician as well as the body, and the moral and mental management of the sick is often far more difficult than the physical. A close, thoughtful study of mental forces and of mental therapeutics in affections connected with the brain and nervous system is one of the necessities that the regular profession is still extremely deficient in. Irregulars often give a placebo, a mysterious mystery, or a useless agent, which unquestioning faith and other mental operations based on hopes of recovery excited in the wondering patient—*potentizes*—and an astonishing cure (?) results.

This is probably the most rational explanation of the fact, that newly discovered therapeutical agents, as bromide of potassium, salicylic acid, etc., make so many more wonderful cures when first heralded as remedies than they effect after they have taken a definite position in the pharmacopœia.

New or novel remedies often aid the cure through mental influences. Many regular physicians prescribe valuable, true remedies, but give them just as they would administer them to a horse or sheep, as if their only duty consisted in telling the sick what drugs to swallow, and seem to despise the aid of faith, hope, and expectation. You must learn, in simple cases, to depend more upon the aid of hygiene, diet, and mental influences, and less on large doses of disturbing medicine, which might allow room for some patients to say that you had almost killed them.

Bear carefully in mind that Drs. Diet, Quiet, Hope, and Faith are four excellent assistants whose aid you should constantly invoke. The Oil of Time and Tincture of Patience are also very useful in some cases, but too slow and inadequate for others, and unless Dr. Dosomething takes the serious case out of Dr. Donothing's, Dr. Tardy's, or Dr. Timorous's hands, they occasionally give one of the confiding kind a wooden overcoat and put him into the hands of Mr. Sexton.

It is bad to let nature take her course when she takes a bad or a wrong one. You will often see her put a curable patient in his coffin because you were called too late to aid and direct her course.

If, at the time, you indicate, to a patient for whom you prescribe an unpalatable medicine, that it will have a bitterish or a saltish taste, or any other unpleasant quality, his mind will be prepared, and it will not seem so objectionable to him as it would were his mind and palate taken by surprise.

If the directions on the bottle indicate what the remedy is for,—for instance, if you have it labeled "to be applied to the injured foot as directed," or "for the pain in the chest," or "for the cough,"—it will tend to give a certain class of patients faith in its being a direct and special remedy, and cause their minds to act with it rather than against it.

Remember that even a highly proper remedy may be pushed too far, or continued too long. Indeed, cases sometimes reach a point at which it is better to stop all medicine temporarily, and rely on hygiene, diet, stimulants, nursing, etc.

You should keep yourself familiar with the ill effects that may arise from the use of the various drugs which you prescribe, in order that you may avoid producing them, or promptly recognize and remedy them if they occur.

Avoid as far as possible the use of medicine that must be taken "through a quill or tube," or that will burst the bottle unless kept "in a cool or dark place;" on which "no water must be taken;" that must be handled with caution, or that must be stopped when the eyelids begin to swell, or when the muscles begin to jerk, or that the druggist must label "Poison;" especially with medicine-haters and skeptics.

Some people will not send for you until they are really ill, for fear you might put them to bed, and thereby keep your carriage at their door; or salivate them, or entail upon them discomfort instead of affording relief. Others will be afraid you will give them quinine, or injure their teeth with iron or calomel, etc., or that if they once begin to take medicine they will not be able to stop. Disabuse the minds of all such people with the assurance that their fears are groundless.

Herb doctors, root doctors, vegetable-pill makers, and others have created a belief in the public mind that remedies obtained from the mineral kingdom—iron, mercury, arsenic, lead, lime, etc.—are poisonous, and should not be taken; while articles from the vegetable are in consequence thereof innocuous and harmless. The truth is,—the most powerful agents—hydrocyanic acid, belladonna, elaterium, croton-oil, lobelia, opium. stramonium, colchicum, digitalis, aconite, strychnia, and a long list of other very active agents—are "Purely Vegetable;" therefore, the announcement, "Purely Vegetable," is but one of the numerous songs of these greedy sirens.

"Iron injures the teeth" is a remark which you will often hear, and it originates in the fact that the old muriated tincture of iron (tinct. ferri chloridi), which contains muriatic acid, if given without the usual caution, will injure the teeth, not on account of the iron, however, but of the acid that is associated with it; just as the water that makes a pot of boiling coffee would scald a person all the same if the coffee were not in it. Preparations of iron containing no free acid do not act upon the teeth.

Iron temporarily blackens the tongue and the stools; and it is well to tell persons this to prevent needless alarm.

It is believed by many that *quinine* gets into the bones, affects sight and hearing, causes dropsy, etc. So firmly do some people believe these things that you will at times have to humor their prejudices, and change to sulphate of cinchonia, compound tincture, or some other preparation of bark, when bark is indicated.

This prejudice depends chiefly on the fact that, being powerful for good, people naturally infer that it must be very strong; hence, also, powerful for evil. It is also often due to the teachings of Irregulars, who seek through it to prejudice the public against regular physicians, while constantly but secretly using it themselves. I have known a conspicuous Irregular to denounce quinine strongly, and yet use three-grain ovoid, gelatin-coated quinine pills under the name of "*Panama Beans*" for the cure of his patient. We know that quinine, when properly used, is really an almost harmless *vegetable* product, which acts on the malarial poison not by great strength, but through its antidotal influence, just as water, an agent harmless enough to drink or bathe in, acts on fire.

One of the most provocative and annoying hardships you will have to endure is the tendency of some people who have suffered protracted sickness to blame you or your medicine for any permanent impairment or persistent lingering symptoms after illnesses, instead of recognizing and acknowledging the fact that they are the real effects (sequelæ) of the disease.

Reproach is often unjustly cast on physicians and on medicine by people living in malarious districts, who sicken with this or that malarial affection, send for a physician, and get well, and might remain so, but, being still surrounded by malaria, they again inhale it and are again poisoned. This they erroneously call "a return," instead of a re-poisoning. Of course, while the laws of His Majesty, King Malaria, remain as they are, you can no more promise future immunity to convalescents with an agued frame who remain in malarious districts than you can promise the anxious sailor that future winds will not again create waves, or the uneasy farmer that recurring frosts will not again nip his exposed plants.

Malarial exhalations from the earth are usually greater at night, and greatest on the still, damp nights of autumn. Malarial affections, therefore, are usually contracted in autumn and at night, but many people are ignorant of the fact that they can also rise, and be caught, during the greater part of the year; they can also be caught in the day-time, and persons should be put on their guard.

It is proper, and your duty, to advise a person to change his abode if necessary for his health, or to relinquish an occupation if it be injurious to him. Also to dissuade him from exciting pursuits, or from striving to amass riches when his health is thereby jeopardized; at the same time, bear in mind that such subjects are both tender and delicate to deal with. Be on your guard, therefore.

Keep yourself well informed in regard to suitable clothing, physical exercise, and proper diet; also as to the value of pure air, pure water, and pure soil, the comparative healthfulness of different regions, the presence or absence of malaria at different seasons and places, etc.; also in regard to the various health trips and summer resorts. Familiarize yourself likewise with the constituents and peculiarities of the various mineral waters, and the special uses of each; with the comparative advantages of seaside and mountain trips, and with the classes of invalids to be benefited by one or the other; also with the various baths— hot, cold, tepid, Russian, Turkish, electric, vapor, etc.—and the comparative advantages of the various hospitals, asylums, sanitaria, retreats, etc.; for such matters belong strictly to the province of medicine, and it is especially desirable that you should understand them, because you are sure to be asked about them, and sure to be ashamed if you cannot answer, and the refined and intelligent inquirer will feel disappointment and distrust of your knowledge; they are subjects that concern the better and more desirable classes of patients, many of whom are semi-invalids, with whom you will often have to make hygiene, medicinal waters, trips, etc., go hand-in-hand with medication; also, remember that your duty does not terminate with the cure of the malady which you are called upon to treat, for in almost every case you should lay down rules and regulations to prevent a relapse, sequelæ, or future attacks.

You cannot be too cautious in advising **exhausted patients** with impaired appetite and weakened digestion, or others far gone with obstinate and dangerous, invincible, or hopeless maladies, to leave their homes and undergo the fatigue and discomforts of travel to the sea-shore, mountains, or other distant places, or to foreign countries, among strangers, in search of health, unless there are sound and good reasons for the belief that the change will be beneficial and that improvement or restoration to health will result. The risk of breathing their last away from home, family, and kindred, or of a return made worse by the inevitable fatigues and exposures of travel, is not to be assumed without full consideration. We sometimes actually hear it hinted that the physician has sent the patient away to get rid of him, or because he did not know how to treat him.

Be chary of sending people from their homes to the crowded wards of hospitals, unless you feel assured that the management is kind, humane, and skillful, for, while hospitals and almshouses are an unspeakable blessing to sick wanderers, to the castaways, the forgotten, and the homeless, they are to a less extent, if at all so, to those who have friends and a place to call home.

"Be it ever so humble, there is no place like home."

To remove a weary and worn invalid from the little spot he calls "home" to a hospital or other asylum for poverty, deprive him of his friends, neighbors, and companions, and all the little endearing sympathies and solaces of domestic life, restrict his freedom by prison discipline and half-way imprisonment, and subject him to the sense of friendlessness that is too apt to seize the mind in the hours of sickness, and to the foul effluvia and diseased emanations that lurk about the wards of large and overcrowded hospitals, and to the risk of rugged indifference on the part of paid, possibly coarse, nurses, and to irksome, humdrum, hospital rules—to bed, to meals, to everything at the sound of the bell, gong, or whistle; to expose him possibly to the public gaze, merely as an object of medical treatment, or for experiment with new remedies, or for the clinical advantage of medical students, designate him by a number, clothe him hospital fashion, and put him on diet prepared at regulation hours by stranger hands that know not his peculiarities or tastes, his likes and dislikes—if he be a person of domestic tastes and sensitive disposition, with a natural attachment to his home and its surroundings, such a change would be most hurtful and injudicious, and could scarcely fail to aggravate his disease. Worse still, if he be carried to a medical-college hospital, with its busy crowd, shuffling feet, wilderness of gazing eyes, and sea of eager faces gathered around while he is used as a subject for the demonstration of his disease, or carried through repeated or prolonged examinations for the education of students.

The belief that taking water or ice is dangerous in fever is still very general. People are wonderfully slow to recognize the fact that water, whether applied externally or sipped and swallowed in small quantities at a time, is one of nature's greatest remedies in fever, especially if the patient have a craving for it.

If a person perspire more during sleep than at any other time it is a sure sign of weakness.

You will often be asked, "Doctor, may the patient eat anything he wishes?" If you think that ordinary food will do him no injury, be careful to answer, "Yes; he can have any *simple* thing he wishes." Were you to say, "He can have *anything*," it would include pickles, radishes, cheese, ham, veal, sausage, and a great many other indigestible things that might injure or kill him; the addition of the adjective *simple* will protect both him and you.

You can have a small, single-page diet-list printed for the use of patients, containing every article of diet in common use, alphabetically arranged, at the top of which you can say, "Every article on this list is forbidden except those that are marked." Supply one to each patient requiring it, and mark or erase from time to time such articles as you deem proper.

When you are busy and wish to make a short visit, **do not** tell the patient so on entering, or exhibit a hurried or abrupt demeanor, but begin promptly to ask the necessary questions in orderly relation, and do not allow the patient time to introduce other subjects, or in any way digress from his case until you have learned all that is necessary. Have neither eyes nor ears for anything except your patient. If the subject of the weather is broached, answer as if you were considering it only in reference to its influence on the patient before you, then go back to his case. Economize time thus; but if your patient is ill, neither allude to your haste nor in any way show that you are in a hurry until you have made your examination and written your prescription. After completing the circle of duty and giving all the attention absolutely required, if you courteously depart forthwith, he will not feel that your haste has caused any inattention to his case, as he would if you had leaped into his room precipitately, thrown down your hat and gloves, dropped into a chair, asked a few hurried, desultory questions, made an incomplete examination, jumped to your diagnosis, scribbled off a prescription, leaped into your carriage, and hurried away.

Unless there is an obvious reason for an opposite course, it is better to avoid all desultory conversation on general subjects at the beginning of your visits.

It will often vex you, when you are busy and time is precious, to be kept waiting below stairs while the people in the sick-room prim and prepare to receive you with as much prudery and tedious ceremony as if the surroundings, rather than the patient, were the object of your visit. Show every one the respect due to sex and rank, but manage at the same time to let such people know that your time is too precious to be wastefully expended, and must be divided somewhat equally among those whom you are attending.

Never assign as a reason for being habitually late in visiting an ill patient that you are overbusy. Every one wants a physician who is in active experience and engrossed in practice, but no one likes to be habitually slighted or crowded out. A case of obstetrics or an accident is, however, deemed by all an acceptable excuse. It is an excellent rule always to let patients know at your visit when they may expect your next visit, and to keep your engagement as near to the time as circumstances will allow. Such a system gives satisfaction and prevents anxiety, and you will then generally find them prepared to see you without detention or flurry.

It is very important always to ask to see the patient's medicine as soon as possible at your visit, and to ascertain by both inspection and inquiry whether it has been taken according to your directions *before* you express any opinion of the patient's progress. If you neglect to do so, you may be caught confidently ascribing improvements to prescriptions that have not been used, or to remedies that have either been thrown out of the window or emptied into the garbage-box, and you will become the victim of a never-to-be-forgotten joke.

School yourself to avoid crude remedies and to cultivate conservative rather than radical ones. Throw gross physic to the dogs. A repute for not being heroic in treatment and not giving much strong medicine is at this time a telling item in a physician's reputation, one that might almost be adopted as a corner-stone. Of course, in cases in which duty actually requires you to act promptly and decisively, or to use powerful remedies heroically, you must not hesitate to take the responsibility and do whatever is right and proper.

Avoid polypharmacy. It is much better to order some single remedy, or a combination of which you know the physiological effect, than to order an indefinite medley on the ancient blunderbuss principle.

It is highly proper, and a duty, to warn people of dangers to the public health, and to devise means to prevent or remove such dangers; also, to teach patients the importance of regular living; proper, careful diet, good water, pure air, effective drainage; also, of the dangers that may follow the sun's rays and evening dews, that they may escape disease and preserve health;

but it is neither just nor wise to teach other than medical students the secrets of our art, nor to familiarize the laity with the drugs you employ. You should especially avoid giving self-sufficient people therapeutical information that they can thereafter resort to and ignore the physician. If you do, they will soon feel brimful of wisdom, become opinionated, and imagine they know as much about medicine as you do, or than all our profession combined, and begin amateur prescribing and neighborly doctoring, and not only take your bread from you, but make hobbies of what you have taught them, and trifle with serious affections until the patient's disease is fatally seated; after which even correct treatment may be interrupted by the undertaker. It is your duty to cheat neither yourself nor other physicians out of legitimate practice by supplying this person or that one with a word-of-mouth pharmacopœia for general use. If compelled to give a person remedies under a simple form, study to do so in such a way as not to increase his self-conceit, and make him feel that he knows enough to practice self-medication and dispense with your services; and use whatever strategy is necessary to prevent him from taking unfair advantage of your prescriptions.

It is unwise to instruct a person with rheumatism, gonorrhœa, ulcers, sore mouth, sprains, or any other affliction, to get five or ten cents' worth of this or that remedy, to mix for himself, unless it be one of the worthy poor; for people are sure to become self-constituted doctors and abuse such instructions, and try to teach others similarly afflicted how to treat themselves.

"Every sore-eyed person is an oculist."

It is better to let such persons have the medicine from your office, or to write a prescription for it, with instructions neither to repeat nor lend.

It is also better to make your analysis of urine, etc., at your own office rather than at the patient's house, and to keep the details of the processes, reagents used, points of differentiation, and other secrets of the art, etc., to yourself; else Keensight and his friends will quickly become oversmart, begin to test for themselves, and think they know more than they really do.

"The silent man has many things in his favor."

In prescribing, and even in talking of medicines, you should use officinal and not popular names, unless there is some special reason for using the vulgar, or a synonym.

Confine your prescriptions to officinal medicines, and to preparations whose formulas are public property, as fully as possible, and do not patronize any of the semi-legitimate pharmaceutical catchpennies (about which you know nothing but what their labels and wrappers tell) that are now flooding our nostrum-ridden land. For instance, if a patient needs beef, let him eat beef, or have beef-soup, or beef-tea, or beef-extract made for him; if he needs wine, order for him a suitable quantity of the kind which you prefer; if he needs iron, prescribe the kind and the dose that you think proper, and thereby prevent making yourself a mere distributor of some enterprising fellow's ready-made "beef, wine, and iron," which cheats the pharmacist out of all chance of exercising his proper profession, and frees him from all mental exertion and all responsibility in the matter, since he and his apprentice boy have nothing to do but serve as dealers, and hand it out to customers, just as the grocer and his clerks hand out salt and sugar, soap and candles, molasses and tobacco.

The same hat cannot fit every head, or the same shoe every foot, neither can the proportion of ingredients in any ready-made combination suit every patient. Indeed, what would cure one might injure or kill another.

"What has cured Sancho might kill Martha."

Let it be your firm resolve never to prescribe a proprietary remedy, or one covered by a trade-mark; it is better to shun all such ready-prepared remedies, whether trade-mark, proprietary, or quack, whether advertised to the profession or to the public, whether the so-called formula and the dose are given or not. If you order A.'s emulsion, B.'s lozenges, C.'s codliver-oil, D.'s pills, and E.'s bitters to patients, they will, by association, soon think that X.'s sarsaparilla, Y.'s buchu, and Z.'s liver regulator also meet with professional approval. Determine that you will not aid any speculator in life and health to "strike a trade" in your families; and chiefly for the reason that *their nostrums do more harm than good;* also for the lesser reason that justice to yourself and to all other members of the profession requires you to avoid prescribing or telling patients of preparations that will enable them subsequently to snap their fingers in your face and renew them as often as they please, and to recommend them to others who treat themselves with them without your aid. A single trade-mark prescription from you may sell twenty or a hundred bottles that you do not prescribe, and for which you get neither credit nor compensation.

Endeavor to have your prescriptions labeled so as to prevent indiscriminate renewal, as well as to prevent mistakes in their administration; when they are very important, be careful to have the name of the patient put on the label, that every one may know whose medicine it is.

Remember this: The very best time to tell a patient not to renew a prescription is while writing it. If you fear it will be renewed against your wish, stop short while writing and explain to him why it will be a good remedy, or make some other true remark about it, but that he must take only one bottle of it, or that it must not be renewed. Your order given at that time will seem to be founded on some motive other than that of protecting your own pecuniary interest, will impress him strongly, and will be invariably obeyed; this is probably the most effective of all plans to prevent prescriptions from being renewed and adopted as a regular resort in similar cases. With this exception, make it a rule neither to talk, listen, nor answer questions while writing prescriptions.

Never write a prescription carelessly. Legibility is the first requirement, neatness the second. Cultivate the habit of scrutinizing everything you write after it is written, to assure yourself that there is neither omission nor error, and sign your name or initials to every prescription, but not until you have satisfied yourself that it is as intended. Mistakes are seldom discovered unless at the moment of their occurrence.

In consultation, the prescription agreed upon should be written by the regular attendant, and, if the consultant is still present, should be offered to him for inspection; but only the regular attendant's name or initials should be signed to it.

A very, very useful rule in many cases is to specify the hours at which medicine is to be taken; thus, if it is to be taken every five hours, instead of writing "a teaspoonful every five hours," write "Take a teaspoonful at seven, twelve, five, and ten o'clock daily," taking care that the specified hours do not interfere with those for nutriment, and be especially careful to give instructions as to whether the patient is to be awakened at night, or from refreshing day-slumber, either for medicine or food.

In giving directions in regard to doses, bear in mind that spoons and drops vary greatly in size. Much trouble and uncertainty can be avoided in cases where medicine will have to be taken for any length of time by getting a graduated medicine-glass, which is both convenient and precise. A minim is a definite quantity, a drop is not; therefore, in prescribing potent fluids, you should order minims instead of drops.

Neither alarm your patients nor their friends, nor risk the dangers of the chloral, opium, or other bad habit being acquired, by allowing the sick to know that they are taking such remedies.

If you instruct a patient how to use the hypodermatic syringe on himself, or to inhale chloroform or ether, or give him cocaine, chloral, opium, alcoholic liquors, or other exhilarating and fascinating agents without discrimination, or to use according to his own judgment, if he have any predisposition toward them he will probably acquire the habit; and if he does, you will surely and *deservedly* incur the blame. The slaves of such blighting habits always cast the blame for their acquired passion, or their withering enslavement, on the physicians who first ordered the drug or stimulant for them, if they have the least ground for doing so.

Hypodermatic medication not only has its place as a valuable remedial agent, but at times becomes really indispensable; it also has various drawbacks that should prevent its indiscriminate use. Among the lesser evils connected with it is that those who are soothed and temporarily comforted by it, or have become habituated to it, are apt to harass and worry you for its application at all hours, day and night; and you will often find it a real hardship, after doing your day's work, to be obliged to go and administer a hypodermatic (night-cap) of morphia to the Rev. Mr. Cantsleep at eight o'clock P.M., to Mrs. Allnerves at nine, to Colonel Bigdrinks at ten, and to Miss Narywink at eleven o'clock, and probably be called from bed again to insert the sleep-giving needle for one or all of them before morning.

Much of such work is not only a hardship but a nuisance. Far better is it for both the patient and yourself that anodynes be administered by the mouth or rectum in such cases, than for you to have all this extra trouble, and at the same time expose him to what may prove, to him, a fatal charm, and, to you, a sorrowful lesson.

# CHAPTER VIII.

> "Call not on Hercules for help; his aid
> Ne'er serves the man who will not serve himself.
> Thine own arm must the conflict meet,
> Thy purpose being the victory."

It does not require the eye of a Newton or the brain of a Bacon to discover that self-reliance and self-possession are capitally important elements of success. Nothing under the sun will cause people to believe in and rely on you more readily and permanently than to see you believe in and rely on yourself. Your own faith will promote faithfulness. Be not arrogant or self-conceited, and exhibit neither rashness nor weakness, but cultivate self-reliance and the power of thinking and acting in the midst of excitement and distracting forces, and endeavor to conceal all doubts, hesitations, uncertainties, self-distrust, and apprehensions as completely as possible.

> "The wise and brave conquer difficulties
> By daring to attempt them."

Never turn your cases over to "*specialists*," but keep them under your own watchful supervision, unless they present features which render it an actual duty to do so. If you distrust your own capacities, shrink and shirk and timidly refer your cases of eye disease to the oculists, your uterine cases to the gynæcologists, ear cases to aurists, surgical to surgeons, nervous affections to neurologists, throat complaints to laryngologists, mental afflictions to alienists, skin diseases to dermatologists, crooked legs and stubbed toes to orthopædists, warts to a manicure, and so on throughout the list of "ologies," you will lessen your own field of activity, and instead of gaining as much experience with one affection as another, and becoming many-sided and armed at all points,—

> "Dexterity comes by experience,"—

you will soon lose all familiarity with the diseases that specialists treat, they will be "out of my line," and you will dwindle and degenerate into a mere distributor of cases,—

"One starts the hare, another bags it,"—

a medical adviser instead of a medical attendant,—advancing everybody's professional and pecuniary interest except your own, aiding them to gain the admiration of the community, and to make reputation and fees out of that which sinks your own individuality, robs your own purse, and throws (little) you into the shade. A good rule is this: Consult, in cases of irreducible strangulated hernia, stone in the bladder, and in all other capital operations, if you yourself are not, a good operator; also, whenever a serious case, whether in head or body, hand or foot, puzzles or defeats your judgment, or proves wholly unmanageable by usual treatment, or is so grave in prognosis as undoubtedly to require broader shoulders than yours to bear the responsibility, either call in a specialist to aid in its management, or, if need be, turn it over entirely to him. If you study all the branches, and keep yourself abreast of the times, such occasions will soon be very rare to you. Timidity, from a want of confidence in one's own merits, and rashness are both bad traits in a physician, but the former is the greater drawback, since every physician's success must be within himself and must come out of himself, and he must not only have knowledge in his head, but he must have it at his fingers' ends and on his tongue's end, and must not only know how to do a thing, but must also believe that he is able to do it.

Whenever you transfer any one from your care to a specialist's, always do so either by a consultation, a letter, or a personal interview with him, so that he may learn directly from you your diagnosis, prognosis, treatment, etc. You will thereby give him the advantage of your knowledge of the case, and also avoid the risk of any injury to your reputation from an apparently radical difference of opinion between him and yourself; it will also secure your graceful retirement from the case. At the same time, be careful to make your patients fully understand that in turning their cases over to a surgeon or specialist you have only turned them over *for that special affection*, and do not cease to be their physician in any future sickness.

Ask for a consultation in all important cases in which singular difficulty, or obscurity, or knotty problems are presented, or where any doubt exists as to the diagnosis; also, when you are in doubt as to the propriety of a surgical operation, and in all cases in which you think either the patient's interest, his lack of improvement, the appearance of fresh or puzzling symptoms, or a division or sharing of the responsibility demands it; for then, another eye, or a different mind, or an older or more experienced hand, may be of great service. When from any cause you see that confidence in you is wavering, or that necessity for a consultation is arising, endeavor to anticipate the family by being the first to propose it.

Do not conclude that a request for a consultation always implies a mistrust of your knowledge and skill, for it is oftener due to the natural anxiety of the patient's family and friends. We often perceive the limits of our own resources, and, sympathizing with the patient's friends as to the result of the illness, are most willing and anxious for a consultation. On the other hand, when they wish a consultation, they should respect your feelings and susceptibilities, and make known their wish to you, instead of taking you by surprise by springing a consultant upon you without notice; they should, indeed, never send for a consulting physician without your express consent, yet, when it is done through their ignorance of ethics, the discourtesy had better be condoned and the consultation held, for were you to decline to join a consultation under such circumstances you would incur a grave responsibility.

Consultations lessen personal responsibility and, in some degree, anxiety. Besides, they are highly profitable to the profession in more ways than one, and conducive to the advantages of the sick. When you chance to have bad surgical and other cases, or an operation in which life will be risked, or difficult or complicated cases of midwifery, or great or anomalous cases of any kind among your personal friends or relatives, or so near home as to involve you personally or socially; or in a neighborhood in which a group of patients is likely to be unfavorably impressed if the result be unfortunate, it is especially necessary and judicious to call in a consulting physician or surgeon to lighten your burden, even though you have him to come for but a single visit,—to satisfy, if for no other reason, the persons concerned.

If possible, always select high-minded, honorable physicians as consultants, who will second your efforts by their skillful knowledge, and at the same time be likely to harmonize with you in the management of your cases; for their kindly sympathy and co-operation may be highly necessary to the welfare of the patient and to your own reputation; but, when a consultation is held merely to satisfy a patient or his friends, it is then better to throw the selection on them, and to accept whoever is offered to you, if he be a regular physician and a gentleman.

When you happen to be the one consulted, do not enter the sick-chamber, or examine the patient before your conferree's arrival, or ask him questions, except in the presence of your conferree, and have all communications in his presence.

Be punctual to the minute in keeping consultation engagements. You have no right to waste another's time in such cases, or to impose upon him the necessity of awaiting idly for you at the place of meeting.

Under ordinary circumstances it rests with the consulting physician and not with the regular attendant to name the hour of meeting.

In your earlier consultations you will often feel no little anxiety and suspense while waiting to see whether the consultant will act fairly toward you and strive to hide your demerits, or whether he will, by nod or wink, hint, question or innuendo, expose your deficiencies to a few, to be told to many, until you are reduced to a mere cipher in the estimation of those to whom the case is related. To the honor of our profession be it said, that the vast majority of its older members are not only punctilious, but really kind on these occasions, as if still remembering the anxieties and responsibilities of their own early professional lives.

A radical change of diagnosis and of treatment, or a reverse and opposite course in any respect, as the result of a first consultation, often and very naturally impresses the laity with the idea that the previous diagnosis or treatment has been either faulty or actually wrong, and, therefore, unless some real necessity demands it, no material change should be proposed or allowed *at that time.* As a rule, the fewer the apparent changes resulting from a first consultation the better for the family attendant, and especially if he be a *young* physician with insecure reputation.

No physician who has the least regard for honor or principle will persist in an error of which he is aware. If you are ever brought into contact with a colleague who, through what appears to be lack of wisdom, or from self-conceit or sinister motives, persists in differing from an opinion or course that you are sure is correct, or insists on doing what you believe to be maltreatment, or shows culpable neglect, and you fear he may injure you thereby, either withdraw, or insist on calling some eminent member of the profession into the conference, that he may decide between you. After that is done, if you think your interests require it, you can retire from the case without discredit.

When a consulting physician or a surgeon is designated and called at your request, you should see that the payment of his fee is not neglected, and you might with propriety broach the subject to those who are to pay the bill, before he quits. This can be done by privately informing them that his charges will probably be somewhat less if paid at his last visit than if they wait for him to send a bill, which might then be for the maximum amount.

To prevent misunderstanding, it is, in many cases, wise to say a word or two about consultant's fees to the patient or his friends at the time the subject of having a consultation is first mentioned.

You can, in such a case, speak much more plainly on behalf of your brother physician called at your instance than you could for yourself. His relations to the case presuppose him to have nothing in view but the welfare of the patient, and to be thinking only of the scientific and therapeutical aspects of the case, and not of his expected fees. Prompt settlement of the consultant's fees will sometimes conduce, moreover, to a more prompt payment of your own.

Never make yourself responsible for the payment of another's fee. Aid him in a proper degree to get it, but do nothing more.

It is, for several reasons, better for the consultant to send his bill before the regular attendant sends his; when the latter sends his first, it looks as if he is more anxious for the safety of his fees, hence is in a greater hurry than the stranger is.

Unless the consultant gets his fee cash after the consultation, or you are aware that special arrangements exist for its payment, be careful to inform the people, as soon as his attendance ceases, or at any rate before the time arrives for sending them *your* bill, whether he will render his bill separately or not. If you neglect to explain this to them, they will almost surely think you ought to pay him out of your charges, and a misunderstanding will result as to whether you or they should pay his bill.

Whenever, to please the patient or his friends, you are forced to set aside other duties in order to meet another physician in consultation, it is right that you should charge twice as much for such service as for an ordinary visit, or perhaps even as much as the consultant does, inasmuch as consultative meetings not only involve extra time, but the carrying out of the details will devolve upon you and entail additional trouble, and consequently you are entitled to extra compensation.

In dispensing with the services of the consultant when no longer necessary, take care to secure his acquiescence, and make him see that it is done with a feeling of amity and good will.

In consultations it is proper for the regular attendant to precede the consultant in entering the patient's room, and to follow in leaving it.

Friends of your stubborn-case patients, who have special confidence in their own physicians, will often persuade, and sometimes convince, them that you do not fully understand their affection, and strenuously advise them to call in their favorite. In such cases remember that you have no right to object to a patient's having the advice of any one whom he particularly desires in addition to your own whenever he insists upon it; but also, that you have a like undoubted right to refuse to consult with an irregular or any one who is antagonistic to the profession, or whose conduct you deem unprofessional, or who is unfitted for the case; also, any one who is highly objectionable to you for any other reason, or in whose keeping you deem your reputation and interests unsafe. If you are attending a case, and such an one is pressed upon you, you have a perfect right to retire, and should at once offer to withdraw, and thus afford your patient the liberty of choice between you and your rival. Fortunately, such dilemmas are very rare.

Do not refuse to consult with foreign physicians, doctresses, colored physicians, or any other regular practitioners; for you, as a physician, hold a quasi-official position in the community, and, in the discharge of your duties, should know nothing of national enmities, race prejudices, political strife, or sectarian differences. RESCUE! is our battle-cry, and you, as a physician, belong to the world of mankind, and have no moral right to turn your back on sick and suffering humanity, by refusing to add your knowledge and skill, on a plane of real and brotherly equality, to that of *any* honorable, liberal-minded person who practices medicine, if his professional acquirements and ethical tenets give him a claim to work in the professional field. It is not only unmanly to make a class distinction and throw obstacles in the path of the less favored, but such a spirit is wholly incompatible with the objects of our profession (which is a liberal one), and at direct variance with the spirit of science (which is cosmopolitan), and in its efforts to diminish suffering and baffle death recognizes neither caste, pride, nor prejudice, and knows no limits except those of truth and duty.

But while you bid "All Hail!" and give the right hand of fellowship to every honorable, unrestricted physician, and become the friend and brother of all the friends of rational medicine, no matter what their misfortunes or how great their deficiencies; you must, on the other hand, remember that medicine is a liberal profession and not a mere trade, and refuse to extend the hand of brotherhood to any one belonging to a party or association whose *exclusive system*, narrow creed, or avowed or notorious hostility to our profession, prevents him from accepting every known fact and employing all useful remedies, whether dug from the earth, taken from the air, or wrested from the sea,—to any one and every one who cannot honestly say his mind is wide open for the reception of all medical truths, and that his hand shall not refuse to use anything and everything under the blue vault of heaven that may be needed to relieve suffering and save the life of a human being; as that constitutes a voluntary disconnection from the profession. When called in to a case in which the medical attendant cannot do this, you cannot agree with him, and must let his retirement be one of the conditions on which you will assume charge.

You may, however, be called to a case of pressing emergency, such as an alarming hæmorrhage, poisoning, drowning, choking, convulsions, or difficult labor, and find on your arrival that an irregular practitioner or quack is in attendance, with whom you are thus brought face to face. In such urgent cases the path of duty is plain, for, owing to the great danger to life, the higher law of humanity will require you temporarily to set aside ethics and etiquette, and to unite your efforts—head, heart, and hand—with those of your chance associate. Treat him with courtesy, but studiously avoid formal consultations, or private professional dealings, or whispering conversation with him, or any other act that might imply association in consultation.

Thus, you see, there is not only no antagonism between medical ethics and humanity, but that they overlook all questions of etiquette, and allow and cover any and every act honestly performed for the benefit of humanity.

Fortunately, the indications for rational treatment are generally so very clear in such cases that no one can ignore them. If the Irregular has assumed charge before your arrival, and is pursuing proper treatment, or assents to the proper treatment suggested by you, that is all you can ask; for instance, if the patient has received a terrible burn, and linseed-oil and lime-water or a strong solution of soda are being applied, or other rational treatment, indorse it, and advise its continuation; but if your accidental colleague is a hydropath, and wrongly insists on a wet-pack because he is morbid on the use of water, or one who obstinately advocates a lotion of cantharides because they burn and blister people in health, it is your duty to your patient, and to yourself also, unyieldingly to insist that a rational course shall be pursued if you are to take part in the case. Be cautious and firm in dealing with such contingencies, and it is a duty which you owe both to yourself and to your profession that you terminate the accidental and unnatural connection—in a gentlemanly way, of course—as soon as the pressing urgency will admit.

Some unreasoning people may think you are illiberal in refusing to fraternize and consult with Irregular practitioners, regardless of the fact that they have voluntarily divorced themselves from the profession and boastingly assumed a name intended to notify the public that their system differs from ours; and, moreover, that they are hostile to it and to us. Bear in mind that our refusal does not arise from a false sense of dignity or from prejudice, but that the great principle which underlies it is this: as lovers of *all* medical *truths*, we have no fixed, no unchangeable creed, but hail with delight every etiological and therapeutical discovery, no matter by whom made, and take by the hand and recognize as a brother *any one* who is liberal enough to consecrate his life's labor to the relief of the sick; but when we know that a certain person, even if he has an armful of diplomas, circumscribes himself to half a truth and practices a botanical system *only*, or, like a pigeon with but one wing, a vitopathic system *only*, or a hydropathic system *only*, or an omniopathic system *only*, or an electropathic system *only*, or any other one-idea system *only*, and is so tied down and limited to that, by his love, bigotry, or prejudice that he *denies* the usefulness of all other known and legitimate means of aiding the sick, and endeavors to poison the public mind against all other therapeutics but his own,—all rational physicians esteem such an one as *too illiberal* to be a true physician, and justly exclude him as unworthy of fellowship with those who profess to love all truth, and, whilst he remains imprisoned within his own ball-and-chain system, themselves endeavor to steadily pursue, with perhaps less zeal, but with more sense, the path of true science and progress.

If, on the other hand, he uses the remedies that rational medicine supplies, yet adopts the cloak of an "ism" simply as an advertising dodge,—

"Blow, blow, bugle blow,"—

to make the public believe that he practices in some manner diametrically opposite to our system, and thereby assists our opponents to lessen public esteem for legitimate medicine and to create aversion to us as its followers, he is guilty of fraud, and you should, therefore, even on the ground of morality, refuse to countenance him.

When people ask you "what system of medicine you practice," you may very properly reply that you are simply a DOCTOR OF MEDICINE, a PHYSICIAN, a member of the regular unrestricted medical profession, that you have no fixed orthodoxy, belong to no sect, and are limited to no "ism," "pathy," or "ology;" that you stand on a broad, UNSECTARIAN platform, and are at liberty to think whatever you may, only seeking to do your best for every sick sufferer who trusts to your skill and honor; that you accordingly try to be *rational*, and, like the bee, take the honey of truth wherever you find it; that as rational, liberal physicians, the regular medical profession, to which you belong, has no branches, no sects, no dogmas, and bears no man's name, for it is simply the work of the human race, and is held together solely by the common bond of rational medicine; that it maintains perfect freedom of opinion and practice, selects any remedy it pleases, in whatever dose it pleases, and under whatever theory it pleases, and, unlike the various "limited schools," has no articles of faith which it imposes on any one, but accepts all truths, whether winnowed from the store-house of centuries, or discovered, either scientifically or empirically, in our own day; and that you, as one of its representatives, stand ready to embrace and utilize any and every valuable discovery, no matter when or by whom made.

"I shall this good lesson keep,
As watchman to my heart."

This freedom and latitude explains why UNRESTRICTED MEDICINE IS ONE OF THE THREE LIBERAL PROFESSIONS, and why the humane and benevolent physician of the body takes rank with the learned expounder of the law and with the worthy man who inculcates religion, all three uniting to protect the interests of soul, body, and estate. Bear proudly in mind, however, that our useful and excellent science is the only one which regards the entire man, physical, intellectual, and moral; for the lawyer looks on a man as a being possessing certain rights, and subject to certain duties to his neighbors, whilst the divine looks on man simply as a moral, responsible being, who has, or should have, a conscience, to which he directs his ministrations.

To this triad of professions was long ago applied the term

"LIBERAL," because for their pursuits, preventing and curing sin, preventing and curing disease, and preventing and curing legal wrong, each of the trio requires the utmost perfection of character, and because the high-souled sons of law, religion, and medicine have in all ages pursued their avocations as freemen, with hands unfettered and tongue untied, subject to no bonds except those of TRUTH; and yet, as if to blur the grandeur of the picture, law has its shysters, religion its hypocrites, and medicine its quacks. If at any time during your career any sect, schism, or one-sided school arise, no matter how great or how humble its pretensions, if it have even one grain of life-saving or health-guiding wheat to its bushel of chaff, it is your duty to seize the grain of wheat, plant it in the domain of rational medicine, and to cast the chaff, brambles, and thistles to the winds. This determination to enlarge our field of knowledge from all possible source is our life-blood, our invincible strength, and our distinction, the saving element that will cause regular, liberal, rational medicine to exist as long as there are sickness and suffering in the world, and the great feature that distinguishes genuine medicine from all "new schools," "isms," and "pathies."

Remember that we have no secrets, no patents, no monopolies; and that our books, our colleges, our laboratories, our lecture-rooms, our medicines, and the door of the profession itself, are open not only to the newly graduated and the regularly initiated, but to every one who has the necessary educational and moral qualifications, even though he may have been an outsider, allied, whether from ignorance or choice, with schools which are antagonistic to the profession; in the latter case it is only necessary for the applicant to drop his distinguishing creed or system, abandon the hostility to the profession which it implies, and to allow ethical rules to govern his conduct; therefore, no conversion, no standard of orthodoxy, no surrender of private opinion or of favorite theories, or hypotheses, or of unlimited freedom to practice as he chooses, is at all necessary, and each and all such should be individually invited to cease to foster irrational, absurd, and credulous doctrines, and embrace true, rational, scientific medicine.

Be religiously exact in everything that relates to consultations. Let them always be conducted in proper form, and strictly private; before entering the patient's room give the consultant a brief sketch of the history and treatment of the case, then invite him in to a chair at the bedside; after making the necessary examination and inquiries, retire and consult, within a room that is private and exempt from intrusion, if possible; exchange thoughts in an undertone, and out of the sight and the hearing of eavesdroppers, and, if possible, never let your conversation be overheard,—

"The very shadows seem to listen,"—

and never allow any one to be present except the physicians engaged in it. When the consultation ends, the attending physician should, of course, re-enter the sick-chamber and give all the directions, etc., determined upon in the consultation.

Bear in mind, also, that consultations are called for the purpose of deciding for the *future*, not to criticise the past; however, if you are called in to a case, and find that the attending physician is suffering unmerited odium for his previous treatment, every principle of honor should impel you to *volunteer* to defend him. Beyond this, never pass any opinion on the plan of treatment in hearing of the patient or friends, unless it be an approbatory one, and where the circumstances truthfully admit of it.

Let all that follows a consultation show that you act in concert and that it is the result of joint action, and never express an individual opinion of a case seen in consultation, except in strict accordance with the ethical code. If you do, those whom you address may, either unintentionally or purposely, misinterpret what you say, or otherwise discreditably involve you.

Remember, moreover, that if you are sufficiently agreed to continue in joint attendance, you are in duty bound to act in concert, uphold each other, and refrain from telling whose opinion prevailed, or by whom the course pursued was suggested, and from all other hints and insinuations likely to diminish confidence in your fellow-attendant.

If for any reason a professional friend ever request you to see a case with him, not so much for the patient's welfare as on his own behalf, *i.e.*, to confirm a correct diagnosis, and thus protect him against undeserved censure, or to divide unusual responsibility toward a poor or worthless, but exacting patient, or to advise what course to pursue under any other very trying circumstances, you should lend him a ready and willing hand, and that, too, without expectation of a fee.

"Hast thou no friend to set thy mind aright?"

It would be sad, indeed, were any honorable physician to fail to find at least one medical friend to consult without fee in such a dilemma.

Be prompt to the minute in answering all professional correspondence.

If you are ever requested by letter, or by a messenger, to prescribe for an out-of-town patient who is not under the care of any other physician, it is perfectly professional to do so, if you wish, even though you may never have seen the case; but, unless the case is a clear one, it might not be judicious.

Revere the past, have confidence in the present, and hope for the future of our glorious profession, and strictly avoid disparaging the individual members of the profession, or the profession itself, or telling people jokingly of the mistakes and discreditable dilemmas of yourself or others; and also avoid decrying and ridiculing medicine to the laity, and boasting of your own and the general ignorance of disease and remedies, and your distrust of your own capacities, or of the number of people killed, maimed, mutilated, or destroyed in health, and suppress all other fulsome confessions.

"Evil is wrought by want of thought,
As well as want of heart."

When a physician makes such unguarded and sweeping

remarks, he means them *relatively* only; he means to say that he is aware and willing to confess that medicine has its natural limits, and is not an exact science, and that the application of therapeutics is but an art. The public, however, cannot appreciate the sense in which such imprudent confessions are made, and they are taken up by Doubting Thomases and Lying Pauls as quickly as a sponge takes up water, and work no little harm to physicians who make them and to the profession at large; because, all who hear or read them conclude, with Tom Hood, that "it takes a great many M.D.s to be worth a d—n," and that our prescriptions are only a series of guess-work, and that medical practice is only a shapeless mass of uncertainties, contradictions, and inconsistencies, as irregular and lawless as that of the winds; whose votaries ask whether we are certain of anything, or certain that we are certain of nothing; and ever after crack jokes at our expense,—

"God cures and the doctors charge,"—

and either do not employ physicians at all, or do so with feelings of disrespect and distrust.

You know there is no such thing practically as a perfectly straight line, plane surface, regular curve, exact sphere, or uniform solid; yet you never hear the engineer or the surveyor boasting of it from the housetop, or in reckless language, as if to belittle his own profession. Look at the other learned professions: law is still very imperfect and full of uncertainties, and it has its reproaching pettifoggers just as we have our quacks. Its books teem with conflicting opinions, and the best decisions of to-day are liable to be overthrown by others of to-morrow. Religion, too, has its opposing creeds, its rival spires, and its innumerable sects, and its ignorant and often unprincipled expounders,—sad proofs that Medicine is not alone imperfect.

"All things are big with jest; nothing that's plain
But may be perverted if thou hast the vein."

The truth is, physicians personally are far more imperfect than physic. For instance, there are undoubtedly medicines the action of which is *diuretic;* but *diuretics* may be given when not indicated, or the *diuretic* given may not be the proper one, or it may be given in improper doses or at wrong intervals, or it may not be continued long enough, or too long, or without proper restrictions. Now, none of these errors are justly chargeable to the class of medicines which we call *diuretics,* nor to the art of medicine, but are plainly due either to the physician's bad judgment or to his ignorance.

"A hand-saw is a useful thing, but not to shave with."

The fact is, all studious physicians are more or less conversant with the same remedies, but skill in effecting a cure with them consists in applying one's knowledge correctly, in thinking of and selecting the proper ones, skill in proportioning the dose, and genius in judging correctly the time and necessity for their use, etc. Just as different persons essaying to paint will exhibit different degrees of success: one possessed of natural aptitude or special gift will obtain wonderful skill, another less apt will reach mediocrity, while a third will fail entirely in his attempts and quit in disgust,—this difference in result being due not to a difference in the material or colors at the command of each, but to the more or less perfect judgment and skill shown by each in selecting and using them. There must be a reason for giving medicines and also for withholding them, and there must be medication in sufficient doses when there is an indication for it.

The ability to determine accurately the condition of a patient, and to conceive and do the right thing for him at the right time, is the essence of skill, constitutes the chief difference between successful and unsuccessful physicians, and explains the reason why the prescriptions of some physicians are much more valuable than those of others. One may know a vast deal about the profession and yet be a very poor practitioner.

A judicious use of medicines, and not a wholesale renunciation of them, is a leading characteristic of a good physician. When you hear of a physician who wishes to be regarded as especially clear, or ahead of others, or exceptionably fair in his opinions, boasting that he is skeptical, " does not believe in drugs," " depends on kitchen physic," " on nature," etc., you can safely conclude that he has a very weak spot somewhere; either that

"He has mistook his calling,"

or in his zeal to become a medical philosopher, or to coquette with somebody else's opinion, he has lapsed in his materia medica, or overstates his credulity, or that his usefulness has run to seed.

Does the mariner lose his faith in navigation because ships are tossed by the winds and waves and too often wrecked by uncontrollable storms? Or does the farmer deny the fertility of the soil because his neighbor has neglected the proper season for planting and the right mode of cultivation? Or does he lose his faith in agriculture because droughts and insects, and irregularities of sun, rain, and frosts, sometimes ruin his crops? Would any worthy sailor fold his arms and do nothing while the storm raged, or any thoughtful farmer neglect to plant again when the season returned, because the sailor's brightest hopes are sometimes crushed and the farmer's fairest prospects are often blighted? Or is medicine to be abandoned because in some cases it is unable to do all that is expected of it?

"Medicine is God's Second Cause of Health."

Is there a physician on earth who would let intermittent and remittent fevers take their course without drugs, or who would let the syphilitic and other poisons develop or progress unchecked? Is there one who, in the face of the positive facts offered by anatomy, and physiology, and pathology, and chemistry, and hygiene, and materia medica, will confess that he can do *nothing* for pain, or for fever, for nervous complaints, for digestive ailments or chest diseases; nothing for diseases of the circulatory system, delirium, insomnia, headache, epilepsy, hysteria, gout, neuralgia, worms, colic, acidity, peritonitis, child-bed fever, constipation, convulsions, diarrhœa, anæmia, scurvy, cholera morbus, poisoning, casualties, etc. ?

The end and aim of medical practice being to relieve, to cure, and to prevent death, if there is a physician in the land who has never seen medicines restore health or prolong life, and does not sincerely believe in his power to benefit by drugs some of the twenty-four hundred diseases and modes of decay to which mankind is subject, he is *an icy infidel in medicine*, and should at once and forever, for conscience's sake and for the sake of the afflicted, take down his sign, burn his diploma, drop his title, and no longer pretend to practice. What think you of a man preaching religion and living by the pulpit who does not believe in the usefulness of religion?

"Seem a saint and play the devil."

The tolerance of disease has greatly increased in the last few decades, and is still increasing, and medical theories and practice are undergoing great changes. The advance of scientific observation is constantly teaching us to distinguish more clearly between the numerous self-limited cases daily met with and the few that threaten a fatal issue, and *of course* we of to-day use much simpler remedies for the former class than did our predecessors; but it is doubtful whether in serious illness we have *lessened* the doses half as much as some imagine. We now give twelve or fifteen grains of quinia daily for an intermittent fever, where physicians formerly gave half an ounce or an ounce of crude bark containing but six or twelve grains. We give to-day the same dose of opium, or its representative, morphia, when that drug is indicated, as they gave a hundred years ago; the same quantity of castor-oil at a dose, and so throughout the whole materia medica. The great difference is, that we do not now prescribe vaguely or rashly, and when cases are trifling, or obscure, or undeveloped, our treatment is tentative instead of heroic.

We of to-day know better than our predecessors the natural history of disease, and are aware of the almost infinite resources of nature, and that three in every ten of those who send for physicians need no positive medication; that recovery from disease is everywhere the rule and death the exception; and that nine of the ten would get well, sooner or later, by proper hygiene, air, exercise, dieting, and intelligent nursing if there were not a drug or a physician in the world, and consequently we are naturally prescribing less and less medicine. In acute affections, and especially the cyclical diseases of children, we now, in many cases, mainly trust to nature, and see them get well spontaneously from seemingly hopeless conditions almost as if by magic, and these cases constitute a majority of those that seem to be restored to rosy health by a thousand and one therapeutical illusions and quack medicines now in vogue.

The deduction to be drawn from these facts is, that the prudent physician may show as much—nay, more—skill in withholding drugs, and especially those of an active, perturbing character, when not needed, as in giving them when they are.

# CHAPTER IX.

"Pledged to no party's arbitrary sway,
Follow Truth, where'er she leads the way."

BEAR in mind that nothing under heaven prevents you from giving whatever you believe to be best for your patient, whether its therapeutic action be similar, antagonistic, or anything else in the circle; but if, in so doing, you adopt a narrow or foolish dogma, or an exclusive system, and prejudice your mind against all other ascertained truths, your one-sided partisanship will fetter you, abridge your usefulness, and make you unfit for fellowship in liberal medicine. Thus, when Vincent Priessnitz, with his wet sheets and water-tub, in trying to build a house of a single brick, shut his eyes to everything but hydropathy; and one-sided John Brown founded Brunonianism on incitability; and Broussais went wild on Inflammation, Gastro-enteritis, the lancet, and leeches; and Rasori overdosed with his system of "Contraria Contraries," and rabidly denounced everything else; and Samuel Thompson, in his exclusivism, threw away everything but herbs, they each ignored a host of important facts for jumbles of vain, useless, and fanciful speculations, and thereby lessened their own usefulness and that of all who follow them.

"For never yet hath one attained
To such perfection, but that time, and place,
And use have brought addition to his knowledge;
Or made correction, or admonished him
That he was ignorant of much which he
Had thought he knew, or led him to reject
What he had once esteemed of highest price."

Thus it is with agriculture, and navigation, and every other human occupation, but medical science, above all others, has no goal,—its greatest law is PROGRESS.

Medicine is neither a perfect nor a stationary science; not a single department of medicine has yet reached scientific exactness, and possibly never will. We, as rational, unrestricted physicians, absolutely free to think and free to act, are striving hard to bring its various branches as *near* to perfection as possible, and are willing to learn medical truth and scientific wisdom wherever they can be found, and there is to-day nothing of any value in any exclusive system that is not taught by the teachers of the regular profession.

<blockquote>"And thus *we* are the true Eclectics."</blockquote>

When "New Schools," schisms, or creeds arise, if they possess any new or valuable truths, or remedies of ascertained merit, no matter how great or how small, whether taken from the animal, vegetable, or mineral kingdom; from sponge, weed, insect, or mineral; product of wilderness, ocean, or prairie, we instantly single them out and incorporate them with the great mass, to swell the records of rational medicine, and press onward; so that the medicine of to-day may be said to be a living, moving, growing array, founded on all its yesterdays.

Irregulars of every kind feel that, to exist, they must be at war with the regular profession, and derisively style us "The Old School," "Allopaths," etc., to make it appear to their partisans that a creed is necessary, and that every physician must bear the trade-mark of some restricted school or petty sect, with wide-apart principles; and that we are merely one of these restricted sects, with hoary dogmas, and an old-fashioned and dilapidated, moss-grown, ivy-covered creed, twenty-four centuries old. That,

<blockquote>"Slaves to rusty rules,"</blockquote>

we have by-gone habits, ancient ways of thinking, and set rules and doctrines, which, though good enough in olden days, are now altogether inadequate and behind the times, and in their declension. Their aim in doing this is, of course, to draw unrestricted physicians on to false ground before the public, and to obtain for their own ism or pathy the honorable distinction and the business advantage of appearing to stand co-equal with us; just as in religion one sect (Pharisees *versus* Sadducees, or

Catholics *versus* Protestants) stands with reference to another, and in politics Republicans stand with Democrats.

Faugh on such nonsense! Who are the heroes of modern medical science? What men are TO-DAY, as ever, bearing forward the flag of medical discovery, and making the star of truth shine over the hill-tops of medical discovery as clear as the noonday sun? Regular, unrestricted physicians! Who are the great authorities, and who hold the most advanced views on anatomy, physiology, pathology, gynæcology, ophthalmology, insanity, bacteriology, etc.? None other than regular physicians! Where stand quacks and irregulars of every kind in the upward track of scientific medical progress? What have they done for science? o o o 0———nothing!

Remember, that it is *not* on account of their therapeutics at all that we object to *exclusive* systems and refuse to fraternize with their followers, but because they deny the usefulness of remedies taken from any source but theirs; assume dogmas and systems that are limited, and decry and denounce all else.

Were you to announce yourself as an anti-botanic, anti-omnipathist, anti-allopathist, anti-eclectic, anti-electropath, anti-hydropath, anti-vitopath, or anti-anything else calculated to produce division, antagonism, or strife in the ranks, it would be unprofessional, and equally as inconsistent with the spirit of scientific medicine as the systems you were opposing, and would abridge your usefulness and render you unworthy of professional fellowship, just as it does others who follow opposing creeds.

Although it is wrong to spend much time and labor in acquiring knowledge of anything that is useless when known, yet it is well to look into the principles of mesmerism, hydropathy, galvano-therapeutics, hypnotism, spiritualism, etc., to enable you to speak of them from personal knowledge, and to checkmate their representatives, who, in their arguments to the laity, make great capital out of *knowing all about the* "*old-school system*," which they, of course, aver does not compare with whatever " new school " they practice.

To limit one's practice to any certain segment of the medical circle is, of course, quite different from limiting one's creed. You have an indisputable right to confine yourself to any specialty or department of medicine you please, but as it is a self-imposed limitation of your sphere you should take care in your signs and cards *simply* to add to your general title the words "Practice limited" to the eye, or to the throat, or to skin diseases, or to whatever else your specialty may be. Such an announcement is honest and professional, and claims nothing more in the way of skill than your M.D. presumes. A sign or card with the words "Practice limited to," etc., is perfectly professional; one that reads "Special attention given to," etc., is not.

Bear in mind that we condemn no system or discovery, ignorantly, on the principle which governs the Indian, who disbelieves in the locomotive and telegraph, or on that by which Galileo was persecuted, which, by the way, was theological, not scientific; neither do we accept anything as a blinded Hindoo devotee does his religion; but, on the contrary, thousands of competent, earnest, fair-minded, truth-loving deep-thinkers and clear-seers all the world over, both in hospital and private practice, with open eyes and alert ears, solely for the purpose of ascertaining the truth for the benefit of medical science and of suffering mankind, and anxious to see new links added to the great chain of therapeutic aids, eagerly, and fairly, and deliberately investigate and test all the alleged important discoveries, plausible theories, and so-called reforms in medicine from A to Z when they arise, and by the conjoined result from a thousand sick-rooms and laboratories give a true common-sense verdict.

"When free from folly, we to wisdom rise."

And it is no more necessary for every succeeding generation, with more useful things to think about, to turn aside and discuss whether cholera first appeared at Jessore or in Bengal, or whether Æsculapius was a real person or only a myth, or to re-sift, re-weigh, and re-judge unreasonable medical vagaries and nonsensical dogmas that have been a hundred times disproved before rejecting them, than it is for every one to study spirit-rappings, jingoism, table-turning, the Book of Mormon, and the ins and outs of all other freaks, frauds, frenzies, jackassical doctrines, therapeutic follies, and theoretical crazes, after thouands and tens of thousands have proven them false.

One of the most amazing of all wonders is that wisdom in the mysteries of the law or in the doctrines of theology, acumen in the sciences, skill in the polished arts, or keenness and culture in other departments of human knowledge, scarcely increases some people's reasoning powers a jot above the Ancient Egyptians in medical matters, or prevents their being led astray by false notions of cures, remedies, specifics, and antidotes.

More than one prominent citizen continues to put knotted red strings around children's necks to cure whooping-cough, or a bag of camphor or of asafœtida to ward off the epidemic; more than one sea-captain carries a potato or a "chunk" of brimstone in his pocket, or wears one stocking wrong side out, to charm away rheumatism; more than one millionaire has vowed that globules of tartar emetic have restored his strength; and such is poor human nature, that many a victim is actually ready solemnly to certify that this, that, or the other worthless quack swindle has saved his life, yea, even after he had stood on the crater of the volcano of death and heard the rustling of the black angel's wings.

> "Hence, Sharper, pitch thy trammel where thou please,
> Thou canst not fail to catch such fish as these."

How any individual can be a wise logician in all else, and yet, as soon as sickness attacks him or his, leave all reason behind, and with open mouth and closed eyes become an easy, almost voluntary prey to shallow quackery, and exhibit the strongest faith in sophistical pretension whose assumptions are glaringly contrary to common sense, is a psychical enigma that almost weakens one's faith in the common sense of half of mankind.

Never hold joint discussions or controversies before the non-professional public, with Irregulars, noisy quacks, or wrong-headed enthusiasts, either through the newspapers or in any other way, no matter how false or shallow their doctrines or oily pretenses are, or how easily their weak arguments are refuted by stronger ones; because such joint discussions and rejoinders, either by speech or pen, with the public as judge, would result in no good, but give Pollywantsacracker & Co. an opportunity to make the noise and clamor they desire, bring them into greater notice, gain for them new partisans, and give them a chance to cloud the truth, raise additional false issues, and pose as martyrs to scientific persecution.

You will occasionally be called again to families who, on account of the blunder or misconduct of some unlucky or incompetent member of the profession, strayed in disgust from regular medicine years ago, when bleeding, etc., were fashionable, who will be surprised to learn that the fashion of medical practice has changed, and that your therapeutics differ very decidedly from those of Professor Oldkind and Doctor Van Winkle, and that you have learned to unlearn many things and have not taken an oath to practice as our great-grandfathers did, and no longer bleed, salivate, and give nauseous drugs indiscriminately. If you are prudent and circumspect, most of these can be permanently reclaimed.

"Then, grasp the skirts of happy chance."

But few of the really sick who are persuaded into giving false and one-idea systems a 'trial become converts; common sense prevents. Therefore be careful not to banter, irritate, or abandon people who are trying an *ism* or a *pathy*, or believe in it *a little*, lest from combating their maybes, questioning their prudence, and forcing argument, you *drive* them into these vagaries permanently. Should one even contend that the earth is three-cornered, or that sugary nonsense has saved his life, or that pumpkins grow on trees, or declare white to be black, or that castor-oil is made of dead men's bones, or that a horse-chestnut and a chestnut horse are one and the same thing, laugh in your sleeve if you must, but do not combat him *too fiercely*, for pride of opinion and determination not to be browbeaten into recantation are unfortunate impulses to arouse, especially in conceited and silly people, who admire their own ingenuity in discovering arguments, and will certainly drive them to take sides against you with energy and zeal, possibly to swear by the error in all the moods and tenses, and thenceforth to injure rational medicine to the full extent of their influence; for,

> "Faith, fanatic faith, once wedded fast
> To some dear falsehood, hugs it to the last."

If, in exposing any delusive or false system, you are careful not to denounce it with too much warmth, as though prompted by prejudice or self-interest, and confine your condemnation strictly to the impersonal abstract subject, showing that you speak your real sentiments from sober reason and conscientious devotion to the truth; and if, moreover, you avoid appearing anxious to hoot down or excite hostility against the individuals who appear to practice it honestly, your reasoning will have a great deal more weight with those whom you address, and with the community, than under the reverse circumstances; for human nature is such that, if a system or creed in medicine be false, unkind or untrue abuse of its representatives will be one of the best ways of commending it to public favor, and, therefore, is what they themselves most heartily desire.

Ours is the age of quackery,—quackery in law, quackery in religion, quackery in medicine, quackery in everything. Medical quackery subsists on credulity, gullibility, and ignorance, and, whenever you have a fair opportunity, it is your duty to expose it, and to save as many as you can from its clutches.

Medical laws that discriminate in favor of the true men of science and integrity, and against the empiric and impostor, are everywhere essential to the public health and the public safety; but many of our States have no medical laws at all, and their common laws do not protect their citizens against imposition or enter into the slightest consideration of the worth, or worthlessness, of various *isms*, *ists*, and *pathies*, but recognize all kinds, regular, irregular, and mongrel, even down to notorious quacks and ignoble impostors, who never saw farther into the human body than the skin, precisely as they do the regular profession; therefore, if you ever occupy an official position under such laws, you will have to recognize certificates of death, vaccination, life-insurance, etc., given by irregulars of every shade, no matter how fictitious their pretensions, or how profoundly ignorant of common medical truths, just as you do those of intelligent, rational, honorable physicians. In a word, you will have to recognize officially every person whom the law recognizes. State medical laws that indiscriminately legalize people of all kinds, of all colors, of both sexes, and of all nations, or give a license to practice to every ignoramus, are impaired to a corresponding extent. What we need is proper laws for the protection of the people,—laws that, while recognizing and protecting the rights of all educated physicians, without regard to their creeds or modes of treatment, would unsparingly uproot and weed out the whole miscellaneous rabble of abortionists, self-commissioned faith-cure pretenders, oxygen quacks, the Street-corner Doctor,—

"From his discourse he should eat nothing but hay,"—

Ambulating Electric Itinerants, Indian Doctors with their wongnim and bunyip, Steam Doctors, Pow-wow—

"Every inch that is not fool is rogue"—

and Root Doctors, with their "passel of yerbs," who neither read nor write, but get their "TOLERABL' SARTIN" LARNING ABOUT RUTES AND YERBS, "by revelation from the LAWD"; also Dr. Squish, the "cullud gemman" who learned to cure the conjured "by 'speriments,"—

"Just befo' de wah,"—

and other mean and soulless swindlers who know as little about a physician's duties as they do about a geometrical icosahedron or the constellations of the heavens, but pretend to answer the unanswerable, and make lying promises to cure the incurable; and all other outlaws who knowingly deceive and defraud.

> "Expunge the whole."

Just laws requiring written examinations upon the fundamentals of Anatomy, Physiology, Chemistry, Surgery, Practice of Medicine, Materia Medica, Therapeutics, Obstetrics, Gynæcology, Pathology, Medical Jurisprudence, and Hygiene, should be enacted and rigidly enforced in every State, instead of weak laws that confound the worthy and the worthless, the skillful and the useless, the educated and the ignorant; and not only compel those who administer them to recognize fool-quacks, illiterate boobies, consummate dunces, and a whole troop of "nat'ral" born, darn'd fools, but also lend respectability before the public to knave-quacks who deserve the cat-o'-nine-tails.

> But when you see that blessed day,
> Then order your ascension robe."

Mixed examining boards are objectionable; better to have separate boards, each consisting of seven members, the Board of Regular Physicians to examine all who wish a license to practice Regular Medicine in the State, the Eclectic, Homœopathic, and other boards to do the same for theirs, as in the present medical law of Maryland.

It would be well and wise if all medical diplomas and official certificates were written in plain English, instead of Latin; then everybody could read and see what each was, and when, why, where, and to whom each was given.

Strange to say, nowadays a section of the public, blinded by the waves of sophistry and swayed by the winds of false sentiment, instead of siding with our opponents when they seem to be right and turning against them when they seem wrong, invariably Ha! Ha! Ha! and, with gross unfairness, side with the "new school" or the quack, or anybody else, whenever a contest arises between them and us.

> "Truth forever on the scaffold,
> Wrong forever on the throne."

Even the press seems to delight in aiming shafts at the regular profession and creating popular sentiment in favor of our enemies, by making invidious comparisons between their modes of practice and ours, telling of their wonderful success and steady growth in public confidence in highly colored terms. Censorious editorials and lampoons are frequently written on our arbitrary exclusiveness, our bigotry, etc.; our bickerings and our disagreements, too, are magnified, and our professional squabbles and disputations are reported in a sensational way, all *apparently* to antagonize and decry us and to cheer on and assist the onsets of struggling Irregulars and advertising quacks, under their false but popular cry of "persecution."

You will find that if a person happens to get better, even of an ordinary affection, under the chance play of an Irregular, or by fool's luck when taking a quack medicine, it attracts general attention and every one will speak of it; whereas, if twenty, equally important, get well under the skillful practice of regular physicians, it is considered quite a matter of course, and scarcely excites a comment.

'Tis said the Chinese are so expert in making much out of little that they live and fatten on what a Caucasian wastes. In the same degree, Irregulars and quacks thrive on the quickening influence of the emotions—expectation, faith, hope, etc.—which we, with our minds fixed on more tangible agents, neglect far more than we should. For proof of the mighty power of the mind over the body, look at the liver-pads, tractors, amulets, charms, and dozens of other humbug agents now in vogue, which the young and old, black and white, educated and illiterate, all kinds, classes, and conditions of people, are praising, almost as if they had fallen from the skies.

Fashion and wealth exert a powerful influence in medical affairs, and, unfortunately, the novelty-seeking portion of the fashionable, wealthy, and influential foster with their influence and patronize with their wealth almost every pathy, ology, and ism in medicine, and make them popular and fashionable, while some of the lower strata stand with eyes, ears, and mouth all open, ready to follow every fashionable foible.

Some Irregulars have this source of *éclat*. Having the humbug element fully developed in them, they, with a look of owlish wisdom, big words, and a jargon of technical terms, magnify what we would call a slight cold, or a quinsy, into a "congestion of the lungs," a "bronchial catarrh," a "touch of pneumonia," "diphtheria," or "post-nasal catarrh"; dignify what we would call a disordered stomach into a "gastric affection," a wind colic into "borborygmus," a wen into a "cancer," etc., for the cure of which hard-named diseases they are duly credited in their statistics and fully paid by their patients, who are thus added to the list of "saved," and the family are fully convinced of that ism's or pathy's remarkable power in those diseases. There is a fellow in our section who works this trick so adroitly that he actually reaps more credit and confidence from mistreating a case that dies therefrom than you would receive from one properly treated that gets well, and reaps more credit and patronage for stopping a chill and fever in seven days than an honest physician would for doing the same in a day or two.

Another reason why Irregulars get cases is, that if a physician grows tired of a case and loses interest, or the patient gets tired of him and loses faith, the family is apt to desire a change of treatment, and, fearing the attendant would become offended were they to dismiss him and employ one of his brethren, they get an Irregular, under the belief that the physician will feel *less hurt* if they dismiss him under the plea of trying "a different system" of doctoring than on any other pretext. Besides: there are fully five times as many regular physicians as there are irregulars, and we naturally get more stubborn cases, and more dissatisfied patients, who turn from our larger number to them, to "try another system," than there are to come from their smaller number to us. We suffer more because we have more to lose.

Irregulars have thus been catching numbers of patients

under the idea that they are "specialists in therapeutics." The advent of our true specialists is very fortunate for us in this respect, as more and more of our stubborn cases now fall into their hands instead of wandering off, and are thereby kept with the regular profession.

"For this relief, much thanks."

Again, a physician is sometimes compelled to tell disagreeable truths, and candidly to give a gloomy or despairing prognosis, and this, on the principle of a drowning man catching at a straw, is apt to make the patient and his friends argue that, as regular medicine offers him no guarantee of safety, they had better transfer the case to some irregular practitioner, or to a noisy quack, who makes great professions and rosy promises.

Another reason why Irregulars have partisans is, that legitimate medicine is unsuited to the peculiarities of some minds, and will never obtain their confidence. Some would almost rather die under the hands of an irregular than to recover under a regular. There is, also, always a sprinkling of extremists and pharisees, long-haired men and short-haired women, of every conceivable type of mind,—

"Nature in her time has framed strange bedfellows,"—

opponents of vaccination, Spiritualists, and other well-known malcontents in every community, who for one cause or other are imbued with dogged antagonism to the regular profession, and the fellow who discards it is their doctor; others believe in vegetable remedies only; the prescriber of herbs is their doctor, etc. All such unite, by affinity, to abet and support classes and systems that practice in opposition to us, and, of course, such demand creates a supply.

You will find that not only in medicine, but on every important subject, when the plain, common sense of a community reaches a conclusion, there are always persons who think they exhibit finer qualities of mind by reaching the opposite conclusion, and will contend bitterly on points where rational doubt is impossible. Other "intelligent enemies" think it evinces great natural acuteness and subtlety of intellect to cling to the opposition, and imagine they thus show superior penetration and sapiency.

Still another reason why Irregulars get patrons is this: they all take care to announce that WE cure by mild means or harmless methods, and not by complicated, painful, or dangerous measures, bloody operations under anæsthetics, or other *dernier ressorts* that science teaches truer physicians to use—against all of which they have, by false assertions and fallacious statistics, aroused much of the existing prejudice and abhorrence.

"Fear has big eyes."

So great, indeed, is the popular dread of what physicians *might do,* that in choosing a medical attendant, the nervous and the timid, the friables and the feebles, who constitute *nine-tenths* of all the sick, are greatly inclined to shun Prof. Sawbones, Dr. Doubledose, Dr. Drastic, Dr. Cutemupalive (with his ostentatious preparations and formidable array of instruments), Dr. Bigpill, Dr. Caustic, and all who treat heroically and enforce rigid discipline, and to seek Prof. Tweedlum, Drs. Golightly, Lamblike, and Silky, who undertake to cure without cutting, and who use moderate or pleasanter, even though less efficient, means.

The rational treatment of disease varies from expectant to heroic, according to the exigencies of each case, and you must learn to distinguish cases in which you can safely depend on nature from those that nature cannot combat, and treat each accordingly; for when you learn to recognize those who need an ounce of medicine and a grain of policy from those who need an ounce of policy and but a grain of medicine, you will have entered upon one of the paths of wisdom, and will make yourself and your profession more useful and more acceptable. When you have a Lah-de-Dah patient, with taste or imagination unusually developed, who needs little or nothing, for mercy's sake don't violate common sense and force upon him some horrible mixture that seems as if made of dead men's skulls, or a bitter infusion, or a large bottle of muriated tincture of iron and quinine, or other medley of nastiness, as if your chief aim were to cause nausea and disgust. Give no one anything stronger or coarser than he actually needs, and leave the balance to nature.

Also, handle all who have highly impressible nervous systems, or sensitive skin, delicate palates, tender throats or treacherous stomachs (and wry faces), so to speak, with kid gloves, and be careful to avoid all useless severities, and to give them as little unpleasant-tasting medicine as possible, and never more than they can bear. The recent great improvements in the forms and palatability of medicines, in addition to your own knowledge of the elegant, offer you splendid opportunities to do this. Keep clear of their prejudices, and offend neither their eyes, their ears, their nostrils, their palates, nor their stomachs, and you will succeed where neglect of these precautions might cause failure. Also, bear constantly in mind that painful operations that fail, or disagreeable medicines used unsuccessfully, if they have given pain or great inconvenience, will injure your reputation and may even cause your dismissal. Give hypochondriacs, dyspeptics, and other Pooo-oo-oo-oo-r Cre-e-eatures who are fond of attention, but not of medicine, small, tasteless, or palatable remedies, and, unless there is a real necessity for it, do not oblige anybody to take medicine before breakfast, or to be aroused for that purpose during the night. With such people make free use of bland elixirs, the fluid extracts, sugar-coated granules, pepsins, emulsions, troches, lozenges, capsules, and other results of artistic elegance and chemical accuracy now kept in every drug-store.

Overdosing, blood-letting, salivating, purging, etc., are now justly unpopular, and ultra-conservative, reconstructive medicines are in vogue. Almost every one is filled with the belief that he is debilitated. Say to the average patient, "You are weak, 'below par,' and need building up," and you will at once see by his countenance that you have struck *his* key-note. So much is this the case that many of the ailing, strongly impressed with this idea, will want you to treat them with tonics and stimulants, even when their condition is such that these medicines are contra-indicated.

Never attempt to force the use of a remedy—mercury, arsenic, iodide of potassium, opium, asafœtida, valerian, etc.—on a person after he has exhibited an idiosyncrasy or a hatred toward it. Also, when possible, change the form of your prescription from pills to powders, or from liquids to capsules, or from sweet to bitter, and *vice versâ*, for those who desire it.

A good plan to pursue with patients who actually need the prolonged benefit of two different medicines, who can not or will not take them at alternating hours every day, is to use one to-day and the other to-morrow; for instance, if a nervine and a tonic are prescribed separately, let them take full doses of the nervine on Monday and full doses of the tonic on Tuesday, nervine on Wednesday, tonic on Thursday, etc. Almost any patient can and will alternate thus without tiring.

The smaller, the more striking the means that seem to produce the desired result, the more surprising does it appear to a patient. It does not seem wonderful to him that he should get better after taking on ounce or a pint of anything, but for relief and improvement to follow a tiny powder, or a pellet, or a tasteless solution, or a morphia granule appears marvelously strange, and commands loud praise.

Instead of being armed with paper and pencil only, carry with you a few well-chosen remedies to be used at night, and on occasions of great or sudden emergency. Above all others, carry a supply of morphia granules or tablets with you constantly, and give a proper number of them in an ounce or two of *hot* water as soon as you reach one of the thousand cases in which great pain is a symptom. By so doing you can often adroitly meet the emergency, relieve the suffering, and show your power over pain before the messenger could get back from the pharmacy with the remedy you would otherwise order. It also often prevents the necessity of writing more than one prescription at a visit.

Rest for the patient is rest for the nurse, and, when all around are broken down and worn out, this is an important consideration. The value of a night's rest to a very ill patient is often incalculable, and to secure this morphia granules are highly valuable, even when they form no essential part of the treatment. You can also use them to give any jaded sufferer an occasional night of delicious visions, or of placid slumber, that will make him wonder what has become of the night.

Morphia granules given thus make a vivid impression in the physician's favor, and do great good, becoming, in fact, almost a perfect substitute for morphia hypodermatically.

Endeavor to please every one's taste and ideas of medicine as far as is compatible with safety, and bear ever in mind that a patient is something more than a mere stomach and body; also, study the various psychological aids, and try to compel the patient to assist mentally in curing his own case. Carefully avoid overdosing, and remember that persons who have been most fond of taking medicine often become surfeited and undergo a complete revulsion against both medicine and physicians. How can we wonder at this, when even too long a continuation of beefsteak, partridge, or other savory food causes disgust, even in well people!

This tendency in the human mind has just now received a wholesale illustration at our expense, and in this way: Satiated and disgusted with crude and overactive measures, a great many misdrugged and overdosed people were wishing for a change, when lo! Samuel Christian Friedrich Hahnemann, of Meissen, Germany, accommodated them with a pseudo-scientific, do-nothing system, resting on a creed composed of one logical and two illogical tenets, which nevertheless, by its apparent simplicity, serves specially to advertise both system and disciple, and to fascinate those who trust to it, without offending either eye, palate, or stomach; depending on nature to do what she can do, while itself supinely allowing cases that she cannot restore to become complicated or chronic, or maybe die preventable deaths.

> "Diseases desperate grown,
> By desperate appliances are relieved,
> Or not at all."

You are, of course, bound by the most sacred obligations to use your best judgment and endeavors for the good of every one who comes under your professional care, but neither the Code of Ethics nor the Code of Honor forbids your sailing before any and every popular breeze, provided you violate no principle of truth or justice. No honest man could compromise a matter of principle, *i.e.*, knowingly quit the right for the wrong, or sell the truth to serve the hour, or for one moment permit policy to sit above honesty; yet it is sometimes very foolish not to compromise a matter of mere policy. In medicine the second-best course sometimes becomes the best because the patient likes it best; and, although you can neither believe nor follow Hahnemann's nonsense and follies,—

> "Your key fits not that lock,"—

you *can* follow *the fashion of the day*, and give to every fastidious or squeamish patient the smallest and most pleasant dose that his safety will permit, and *can* avoid giving any one crude remedies to a disgusting degree.

So strong has been the reaction against old-timed medication, that Hahnemann's silly system has secured a large and earnest following, and enjoys the sunshine of popular favor among the susceptible to such an astonishing degree that it may be regarded as the grand delusion of the nineteenth century,—

> "How long, O Lord, how long!"—

and there is to-day no (lawful) human occupation that yields so large a return for the amount of capital and brains required as the practice of homœopathy, and that so few have deserted the crowded paths of rational medicine to seek its shekels is a monument to man's preference of the true and noble path. There can be no doubt that its prosperity would have already terminated had the profession not been so slow to accommodate itself to the demands of fashion, particularly with reference to medication in slight and imaginary cases. But rational physicians are arousing to the importance of this feature, and are rapidly conforming to it. They are also administering more concentrated and palatable forms of medicines in serious cases, and, thanks to the labors of many devoted workers in the field of medical science, and to the light they have shed upon the subject, are now enabled to effect cures with greater certainty, promptness, and safety than ever before. The result is that many of the erring, who had gone over to "isms" and "pathies," are being brought back from delusions to renewed faith in legitimate and rational methods of practice. Determine that you will bear your share in the good work by devoting time and study to rendering therapeutics useful and at the same time cheap, pleasant, and acceptable to patients. If you will carry a small pocket vial-case of your favorite pills, tablets, granules, etc., both strong and weak, for use on suitable occasions, you can meet homœopaths in the matter of free-dispensing, and also have as much benefit as they of the mystery that envelops the name and nature of the drugs thus employed. Besides, you will escape the drug-store "repeats," and if there be any repeating you will do it yourself.

Now, although homœopathy is somewhat fashionable, when a disease actually requires medication you can make little if any rational use of its so-called principles, which rest on the following foolish postulates, which are no more applicable to the treatment of disease than to building a steam-boat: 1*st*. *Curative remedies for the sick can be selected only by a study of provings on persons in health.* 2*d*. *Every remedy must be given by itself.* 3*d*. *The similar and single remedy must be given in its minimum dose*, i.e., *the smallest dose sufficient to effect a cure in the case.* These constitute Samuel Christian Friedrich Hahnemann's substitute for rational therapeutics,—his entire stock in trade; an *essential* triune, an *inseparable* unit,—violation of any one of which is a confessed rejection of this German dreamer's whole system, and you will observe at a glance that it is actually two-thirds nonsense; that the first and second postulates of his creed are sophistical and untrue, and hence should be rejected; and that the last, *i.e.*, to give the smallest dose that will answer the purpose, nobody denies,—

"It is as old as the itch,"—

since it is useless to pour two buckets of water on a fire when sure that one will put it out.

Contrary, however, to what many unthinking people believe, this creed gives its disciples perfect liberty to give either an atom or an ounce of mercury, sugar, opium, or anything else, at a dose, provided they proceed on the so-called principle of similars; and the *question* whether any one does or does not practice homœopathically *does not* at all depend upon the size of the doses. They might give an ounce of a medicine where you would give but a grain. Their ounce would not make them regular physicians, nor your grain make you a homœopathist, for you in selecting your remedies would not think of pathies at all, while they would think of nothing else.

Bear in mind that the practice of rational medicine is also as distinct and free from allopathy and all other pathies as America is from Asia. Here is the true and only test as to whether you are practicing rationally, homœopathically, or allopathically: Were you to examine a patient and ask yourself, without regard to nonsensical pathies, or to any other creed or boundary, *What is the best treatment known to the world for a case like the one before me?* and use that, you would be practicing *rational*, unrestricted, regular medicine. If, on the contrary, you were to examine a patient (with the Will-o'-the-wisp idea of homœopathy in your mind), and ask yourself, *What article would produce a totality of symptoms similar to his in a well person?* and give him the one which you thought would come nearest to doing this, you would be practicing homœopathically; or were you to sit down (with the chimera of allopathy in your mind), and ask yourself what article would jingle with another sympathy or totality of symptoms dissimilar to these, and irrationally base your treatment on that ground, you would be practicing 'allopathically. Now, it is safe to conclude that if you practice medicine forty years you will never sit down by a patient's bedside (conjure the pathies) and ask yourself either "What agent would produce a disease similar to this, or symptoms similar to these in a well person?" or "What would cure by agreeing with dissimilar sympathies?" and attempt to simulate this in your treatment. Therefore, remember that, no matter how small your dose, or what the article, or by whom first used as a medicine, it would not be given by the square and compass of the so-called pathies at all, and you would be proceeding neither *homœopathically* nor allopathically, but rationally.

It is also safe to predict that while reason remains your mistress you can never agree that twelve twelves make a hundred and forty-five, or follow a system of symptom-worship that, in dogmatically seeking to follow the (so-called) law of *similars*, arrives at poison oak-globules as a remedy for erysipelas, croton-oil globules as a remedy for cholera infantum, mercury globules for mumps, tartar-emetic globules for typhoid pneumonia, opium globules for apoplexy, strychnia globules for convulsions, cantharides globules for burns, and an immense farrago of other nonsense as true as, but no truer than—

> "There was a man in our town, and he was wondrous wise,
> He jumped into a bramble-bush and scratched out both his eyes.
> And when he saw his eyes were out, with all his might and main,
> He jumped into another bush and scratched them in again."

Study the "Organon of Medicine," by Samuel Hahnemann; "Homœopathy Fairly Represented," by Henderson; Hull's "Jahr"; Hughes's "Pharmaco-Dynamics"; Johnson's "Therapeutic Key"; the works of Hering, Lippe, and Guernsey, and you will read that a homœopath must prescribe according to what he claims to be the homœopathic, the SOLE, law of nature in therapeutics, comprehended in the phrase "*similia similibus curantur*," or like cures like. It is this SOLE-LAW pretension and false claim to an exclusive possession of therapeutical truth that stamp homœopathy a variety of quackery.

> "Vaulting ambition, that o'erleaps itself,
> And falls on the other side."

It is universally admitted that quinia cures intermittent fever, but who ever heard of its being homœopathic to the periodic feature of that disease? And yet where is the homœopath who does not, in periodic fevers, administer sugar-coated quinine granules in full doses? Podophyllin is the favorite cathartic of the homœopath. Does it ever cause constipation except by the secondary exhaustion and impairment of sensibility common to all cathartics? No intelligent physician would contend that it did. That morphia relieves pain is one of the best-attested facts of medical observation; will any homœopath dare say that it causes pain? These are not stray assertions; they are monumental facts, destined to overturn homœopathy and its silly law of similars; for neither S. C. F. H.'s nor any one else's pseudology can maintain itself permanently before the light of truth and science.

Truth is a unit; there can be but one science of one subject, and there is but one science of medicine, and to talk of rival systems of medicine is as absurd as to talk of rival systems of mathematics or rival laws of gravity. Compare S. C. F. H. (whose whole life was full of unnatural thoughts, foolish ideas, and peevish fancies) and the shallow and delusive so-called sole law of cure, published in his "Organon," in 1810, in which he rails at the profession and talks as if he alone had charge of the key of knowledge and the casket of truth, all the way through, with Copernicus, Newton, Harvey, Davy, Galileo, Franklin, and other real discoverers of nature's laws, who are an honor to the human race, and you will find his baseless and unscientific chain of assertions so weak that

> "Whatever link we strike,
> Tenth or ten thousandth, breaks the chain alike."

Those illustrious men did discover the natural laws of astronomy, gravitation, electricity, etc.; consequently, their systems have extended or have been but slightly changed.

"Truth is God's own daughter."

Hahnemann *did not* discover the natural, the sole, the universal law of medicine; therefore his frail, temporary system has beaten about from psora and dynamization to the tasteless, the infinitesimal, etc., until to-day it scarcely exists except in name.

Among S. C. F. H.'s chief doctrines was the dynamization (spiritualization) of medicines, and his angry assertion that a millionth of a grain of medicine had more power than a grain, or that a drop of alcohol well shaken had more power than an unshaken pint, goes far to illustrate his unparalleled assurance and egotism.

"Yet still some wondered—and the wonder grew—how one small head could carry all he knew."

These notions of his not only contradict reason and violate common sense, but conflict with fixed mathematical laws, since a part cannot be greater than the whole. The truth is, the so-called dynamization, or attenuation, or spiritualization of medicines is bosh, and bears about as much relation to the science of medicine as the kaleidoscope does to the science of astronomy. Indeed, were some graceless wag to exchange or mix up the contents of a disciple's satchel so that each vial would contain attenuations the very opposite of its label, its owner, in blissful ignorance of the fact, would doubtless continue to hear of their great usefulness all the same.

To test the value of dynamization the Milwaukee Academy of Medicine, in 1878, made the following offer: a vial of sugar pellets, moistened with the 30th attenuation of a drug and placed among a number of vials of sugar pellets moistened with alcohol only, to be given to each believer in dynamization found willing to use them, he, at the end of one year, to designate by their effects on his patients which of the vials contained the medicated pellets.

## HIS REPUTATION AND SUCCESS. 249

The project was indorsed by the leading journals and by The New York Homœopathic Medical Society. A mixed committee of believers and unbelievers, of which Rev. Geo. T. Ladd, Professor in Bowdoin College, Maine, was the Chairman, dispensed the sets of pellets, giving them to none but avowed believers in the efficacy of attenuations, each applicant being allowed to name the drug he would use in the trial.

The result was as follows: Number of trial sets applied for, seventy-two; number who ventured to report at the end of the year, ten; number who found the medicated vial, one.

Since this we have heard less and less of dynamization. Hear Hahnemann himself on the subject:—

"It holds good, and will continue to hold good, as a homœopathic therapeutic maxim, not to be refuted by any experience in the world, that the best dose of the properly-selected remedy is always the very smallest one in one of the high dynamizations, as well for chronic as for acute diseases."

Remember that the epithet or by-word "Allopath" is a false nick-name—
*"A thing devised by the enemy"*—
not chosen by regular physicians at all, but coined for us, and put in use against us, by our enemy, S. C. F. Hahnemann, in contradistinction to his own dreamy system, to prove that the theory and therapeutics he proposed in his absurd "allopathy's" place was of a totally different or opposite character from it; and cunningly used as part of his ridiculous attempt to blot out all the existing facts of free therapeutic science, and to substitute his own silly system, and now applied to us opprobriously, with all the collateral insinuations and derisive use the term affords, by all our rivals and enemies, with intent to injure.
*"As the old bird sings, the young ones twitter."*

It is both untrue and offensive, and is no more accepted by us than the term "Heretic" is accepted by Protestants, "Ch—t-killers" by the Jews, or "Locofoco" by the Democrats.

Bear in mind that *we do not study the so-called pathies at all;* therefore, are not "paths" of any kind, but are Rational, Unrestricted Physicians, and take care resolutely and promptly to resent the term "Allopath" when any one applies it to you through enmity, and courteously to disown it, and tell of its falsity, hostile origin, and sinister intent, when applied by those who do not know what malice the term implies.

S. C. F. H.'s so-called allopathic physician would be one whose silly creed tied him to a jargon of pathies and confined him to fiddling on what dreamers call "opposite sympathies," even trying in practice to create some dissimilar, perhaps worse, disease, as a substitute for the one he was called upon to treat. Now, there are fully a hundred times as many squinting-brained people in every community as there are squint-eyed persons, and any man, with but one eye even, can see why policy and self-interest bring forth Protestant bakers and Catholic gardeners, Baptist washerwomen and Quaker boot-blacks, Mormon blacksmiths and Presbyterian shoe-makers, Masonic carpenters and Odd Fellow bricklayers, Republican barbers and Democratic tailors, Methodist-Episcopal astronomers, Homœopathic doctors, French mathematicians, and Botanical druggists, whenever and wherever there are squinting-brained people who prefer to employ dogmatists in these things. Verily it seems as if

"All the world's a stage,
And men and women merely players;"

but everybody knows that no sensible person wants a (Heteropathic) physician who will cure one disease or symptom by creating some contrary, perhaps worse, one; consequently, no thinking person professes to be an Allopath, and no unrestricted physician should allow himself, or the regular profession which he represents, to be thus belied and belittled; but there is, and ever will be, everywhere, a dozen times greater demand for unrestricted, rational practitioners of medicine, who will do whatever under heaven seems best for their patients, without regard to likes or contraries, creeds or pathies, than for one of any other kind. Away, then, at once and forever, with the absurd and false title "Allopath!" and, if a designating title becomes necessary, let it be Regular or Unrestricted Physician.

Also, in signing certificates for life-insurance or beneficial societies, or in giving your name for directories, State or city registers of physicians, or in other cases in which the form requires you to state what school of medicine you practice, be careful to have your name mentioned as a *regular* or *rational* physician, or simply "physician," and not as an allopathist.

Folly-engendering homœopathy not only panders to the whims of the whimsical, but also makes a specialty of poisoning their minds, as well against clearly rational remedies as against the lancet, polypharmacy, and other debatable or spoliative measures, and inclines them to attach undue importance to every trivial affection, and to overestimate the value of placebo treatment, and thus not only tends to produce an effeminate type of patients, who cowardly shrink from every ailment, but also creates a pathophobic overattention to the minutiæ of health, and eventually makes them morbidly anxious about every function, and fills their mind with a medley of imaginary and exaggerated afflictions, which haunt them, like Banquo's ghost, wherever they go.

> "In form a man,
> With spirit less than infancy,
> And nerveless as the weakest woman."

You will often see persons who might have passed through life well and strong, with scarcely a thought of sickness, who, from being indoctrinated in it, softened by its follies, and habituated to its self-surveillance and constant contemplation of their symptoms, become borne down into valetudinarianism or hypochondriasis by a net-work of magnified trifles, and constant indications for pellets and attenuations.

> "I'm doing this for my health,
> I'm doing this for my health,
> For my health, for my health,
> I'm doing this for my health."

We have a wealthy but very fanciful lady living in our section who has become so enamored of similia, etc., that, besides incessantly hunting up indications and symptoms in herself, and dosing herself with globules of table-salt when she dreams there are robbers in the house, of veratrum when she dreams she eats her shoes, and of hyoscyamus when she dreams she climbs the stove-pipe, she also banquets her birds on globules of sun-dew, etc., whenever they fail to sing, and Tabby and Tommy when they fail to mew. Other silly but zealous Hahnemaniacs, as if to complete the absurdity and show their zeal and childlike fascination, have given its similars to turkeys, dogs, chickens, horses, geese, etc., and would almost ask you to believe that they had seen it change a red calf with blue eyes into a blue calf with red eyes.

> "O dark, dark, dark ; total eclipse,
> Amid the blaze of noon."

Not only this, but read the provings in any standard homœopathic work,—Hughes's "Pharmaco-Dynamics," for instance,—and you will find that many of their remedies and many of the symptoms said to follow homœopathic provings are too nasty to be repeated, and that the majority of the others are more like the idle fancies of Spiritualists or day-dreamers than the work of rational persons.

You will discover that nine out of ten of those who to-day long for its sweet nothings know absolutely naught about the so-called principles and sophistical calculations of homœopathy, BUT (and this is a big but) take themselves and give their sugar-plum Materia Medica to their families solely because their remedies are fashionable, novel, easy to take, and prevent the trouble and expense of running to the druggist with prescriptions.

Homœopathy has also profited, and is still profiting, wherever the English language is spoken, by the accidental misleading resemblance of the term homœ to the precious word home,—" Home, sweet home."

To you, as a physician, the term homœopath naturally signifies a person who practices a certain silly, dogmatic, and visionary system. But to many of the laity, on the contrary, the first two syllables of the word suggest that he practices a simple *home* or domestic system of medicine, and the fact that he ordinarily prepares his own globules, solutions, etc., either at his own home or at the homes of those who employ him, instead of sending prescriptions to drug-stores as we do, adds strength to this popular error, which is so natural that many people often actually call regular physicians who happen to supply their own medicines " Home-o-paths."

It is your duty, in the interest of truth and for the benefit of humanity, to make it known that the word "home" is of Saxon derivation, whereas the prefix *homœo* is derived from the Greek *homoios* (similar), and has no possible relation to hearth and home. S. C. F. H. seems to have built better than he knew, as far as the home-loving, English-speaking Americans are concerned, when he styled himself a homœopathist, and not a pathomœist, which has the same meaning.

Do not infer that no homœopath (or omnipath, or hydropath, or vitopath, or any other path or ist) can be influenced by the very quintessence of high motives, or be following his *pathy* with the full purity of truth and the perfect honorableness of honesty, and with the sincerest intention of giving the utmost assistance and relief that the art of medicine enables one to give; for there never has been an absurdity in regard to religious, political, or medical questions that has not found very sincere, well-meaning supporters, educated, refined, and zealously in earnest, who have somehow or other been led into those paths; nor that a homœopath does no good, for he may do a great deal of good, and even get patients well, but the good he does is *not* by his silly pathies, as has been proven by innumerable observers, but by the accompanying tact of tongue, hygienics, dietetics, faith, expectation, good nursing, time, etc., which would do equally much were the *pathy* portion left out, and globules of sugar, saw-dust pills, colored water, or any other make-believe remedy substituted, to give the confiding room to exercise their faith, and rest, regimen, the vital force, time, etc., an opportunity to make the cure.

A preacher, a lawyer, or an agriculturalist may be a fraud; or a navigator, an astronomer, or a mathematician may be a swindler, while the system by which they pretend to be guided is perfectly true. But homœopathy is the reverse of all these; it itself is nothing but a fanciful pseudo-science; hence, not solely the disciple, but the system itself sins against truth.

If ever chance, or a crossing of paths, bring you in contact with a real homœopathist, if you believe him to be a gentleman, harbor no ungenerous feelings or personal animosity, and observe all the forms of politeness toward him, and treat him exactly as you would any other gentleman,—

His faith may be as sincere as yours,—

but ignore him (as he probably will you) *professionally*, and make no attempt to fraternize with him in the management of a case, it being far better for each kind to consult with its own.

Suppose you were to attempt to consult: patient has—well, say, convulsions, the result of teething. You examine the case together—retire for consultation—the subject of treatment is finally reached. You (true to humanity) survey the whole field of rational therapeutics and conclude: first, that the cause should be removed as far as possible by incising the gums for the purpose of severing their irritated nerves; second, that sedatives and antispasmodics are indicated. He (true to his creed) puts on his homœopathic spectacles, surveys *the totality of symptoms* by the square and compass of Similia Similibus Curantur, and arrives at strychnia, in the tenth dilution! Result: emphatically a therapeutic dead-lock, unless, false to your convictions, false to your profession, and false to the interests of humanity, You agree to give up common sense for his nonsense.

But how, oh, how! can any true man have much to do

with any of the other flock,—the degenerate knaves who are not homœopaths at all, but solely for business reasons burlesque as homœopaths and carry awe-inspiring satchels of Hahnemannic nonsense which they handle as carefully, in walking the streets, in getting in and out of their buggies, in going up stairs, and in opening and closing them in the sick-room, as if an additional (here is where the biggest laugh comes in) shake of the powerful dynamizations within might still further increase their terrific potency and cause an explosion.

He seems as far from fraud as heaven from earth.

The whole life of such a fellow, though he lives in luxury and rolls in a gilded chariot, is a living lie,—a sad burlesque on physic, a long-drawn ode to finesse.

Respect every believer in anything, no matter how great his error, if his views be honestly held, provided he show his true colors and fight a fair battle for it; but let the finger of scorn point at every so-called son of Hahnemann who, as an advertisement of himself and to catch patients, denounces and sneers at "Allopaths" and "*old-school*" remedies, meanwhile giving pellets and attenuations *in placebo cases only*, and in all others slyly using our opium to relieve pain, our chloral to induce sleep, our quinia and antipyrin to arrest fever, and all *our* other prominent agents, *just as we do*, in full doses, yet crediting the good they do to *similia similibus curantur*, because just now to call one's self a homœopathist and dispense pellets free to all who will pay for advice, and to surround the name and nature of the remedies with mystery, brings grist to one's mill and shekels to his coffers; but,

"What soul would in such a carcass dwell?"

There is also another variety of fellows, who talk homœopathy to one person and anything-you-wish to the next. These, although inconsistent with themselves, are not so positively dishonorable, for they are at least outspoken in their doubleness; but what words would exaggerate the meanness of a clergyman whose love of gold and lack of scruple would allow him to vary his principles *at will* and preach *anything* one wished, whether a strictly Catholic lecture, or an ultra-Protestant discourse, an orthodox Hebrew sermon, a fiery Mohammedan philippic, or an out-and-out infidel harangue? He might believe in one or none, but he could not believe in all, and, if he professed to do so, would deserve to be kicked out of his own door.

> "An eagle's life
> Is worth a world of crows."

Science has everywhere convincingly shown that Homœopathy contains its own refutation, and is a fraud on science. Hahnemann started it in 1790, six years before Jenner vaccinated James Phipps, and his "Organon" was published in 1810. Now, in this long, long period, had it deserved scientific recognition, or had there been anything at all in it worthy of adoption by the profession, it would surely, like vaccination, electricity, and all other truths, long since have been absorbed by scientific rational medicine, whereas the fact is that its silly creed has taken no root at all in the regular profession. To-day pure homœopathy is withering like a girdled fig-tree in Europe, and Hahnemann is no longer placed on a lofty pedestal—

> "Take him for all in all,
> We scarce shall look upon his like again"—

or worshiped as a hero, and his silly system has almost faded out in the land of its birth, and is without a chair in any university in Europe, and is rarely mentioned there without a smile.

The Homœopathic Medical Directory, recently published in London, shows that there were in Great Britain and Ireland, in 1875, two hundred and sixty-nine homœopaths, and in 1889 but two hundred and fifty-six. In Austria there are now seven thousand one hundred and eighty-three physicians, of whom but one hundred and eighteen claim any connection with homœopathy, and only forty-four of these profess to practice it exclusively; Germany was its birth-place, yet there are but nineteen in all Vienna; and in all Europe, with a population of at least three hundred millions of souls, there are now but one thousand and twenty-two so-called homœopaths,—an actual decrease, and that, too, in the face of an enormous increase of the number of regular physicians.

Even here, in Free America, it has passed its zenith, its feast is almost over, and its MENE, MENE, TEKEL is already on the wall, for it has almost ceased to be the topic of conversation in fashionable circles; the homœopathic book and case of numbered globules are also fast disappearing from the hands of the laity; its so-called specifics are kept among the patent medicines in every drug-store; the system is no longer exciting amateurs, and, far above all, its wholesale desertion of its own principles is about to furnish another proof that no religious creed, no political doctrine, no medical ism or pathy, can extend beyond a limited sphere or period, if it be opposed to the common sense of mankind.

The New York Homœopathic Medical Society and other homœopathic bodies now merely consider Similia Similibus Curantur *the best general guide* in selecting remedies, BUT from which any may depart, if experience or his individual judgment so direct.

Well! well!! well!!! Shades of departing greatness.

Homœopathy has temporarily maintained its ground here not from any intrinsic merit, but chiefly because its pellets, etc., are easily gotten and at small cost, and easily applied in self-medication at the home or from the pocket of its votaries.

In considering the nothingness of homœopathy, think what it is to-day,—that its disciples no longer follow their master, that every change they make is toward regular practice, and that the vast majority of its representatives seem to be on their way back to Regular practice in everything except in name and affiliation.

When Hahnemann started homœopathy, the sciences upon which modern medicine is based—chemistry, physiology, pathology, etc.—had, practically, no existence, and it was this and the overmedication of former days that gave it the start. Were any one to originate such a silly system to-day there could be no excuse for its existence, and it would either fall still-born or be laughed to death.

Study the homœopathic creed closely, and then carefully watch the practice of all those who to-day claim to practice under it, of whom you have personal knowledge, and you will soon discover that few (if any) honestly do so; and although the number of those who pretend to practice homœopathy may still be somewhat on the increase in this country, and enthusiasts here and there are still donating and bequeathing money and holding fairs for the benefit of homœopathic colleges, hospitals, dispensaries, and societies, and its disciples are boasting over their numerical increase and exulting over this *political* favor they have secured, and over that *influential* patient they have netted, and over the other fresh partisan who is praising it, just as they were in Europe thirty or thirty-five years ago; yet, pure homœopathy itself—similia, etc., based on provings—is rapidly disappearing, and I sincerely doubt whether there are at this time half a dozen omnibus-loads of true Hahnemannic homœopaths in our land, and for the confirmation of this assertion I refer to any qualified pharmacist or manufacturer of granules, tablets, and pills who comes in contact with the homœopathic therapeutics of to-day. The genuine homœopath never prescribes tonics, never orders mineral waters, never gives emulsions, never alternates or mixes remedies, and never uses hypodermatic injections, purgatives, mustard plasters, ointments, lotions, washes, liniments, medicated injections, cauterizations, sprays, or gargles; and whoever does so is under the bitter anathemas of Hahnemann, who said: "He who does not walk on exactly the same line with me, who diverges if it be but the breadth of a straw, to the right or to the left, is an apostate and a traitor, and with him I will have nothing to do."

"More might be said hereof to make a proof,
Yet more to say were more than is enough."

## CHAPTER X.

"Behold, how good and how pleasant *it is* for brethren to dwell together in unity!"

Be just and friendly toward every worthy pharmacist. Owing to the relationship and mutual dependence between pharmacy and medical practice, the pharmacists are your natural allies, and should receive your respectful regard. Probably all physicians will agree that in the ranks of no occupation can a greater proportion of gentlemen and manly men be found than in the pharmaceutical. This, and your joint interests, should make you brothers.

It will be found an excellent rule strictly to avoid favoritisms and antagonisms, and to let all reliable pharmacists compete for your prescriptions and for the family patronage which they influence. You will make a serious mistake, and engender active enemies, too, if you go out of your way and without just cause instruct patients to obtain their medicines from any particular pharmacy; if a prescription be properly compounded it makes but little difference by whom, so the compounder is honorable and reliable.

Do not deter your patients from patronizing a pharmacist simply because he is also a graduate in medicine, unless he be uniting the two callings from mercenary motives, or habitually prescribe, or have a drug-store (with a window full of bottles of colored water and quack placards) merely as a stepping-stone to get acquaintances and an introduction preliminary to making his *début* as your antagonist or rival; or if you fold your arms and allow your prescriptions to be compounded by a drug-store physician who *prescribes* over his counter, or in office or parlor, free of charge, and makes it up on the medicine ordered, you will, unless he shows less than the usual amount of selfishness, be apt finally to regret it.

Independently of all other considerations, the joint practice of scientific pharmacy and modern medicine is too much for the grasp of any one human intellect, and a person needs all his time to do justice to either, else one or the other is apt to be slighted; and if your prescription fall into the hands of such parties, or be left to their apprentices or assistants, both you and your patient must take a great many risks.

There is not the slightest wrong in having your name printed on your prescription blanks. But do not use a prescription paper which has any other name upon it besides your own. If it contain the name of a neighboring pharmacist, it will naturally suggest collusion or something else not complimentary; if it contain some enterprising fellow's commercial puff, it will indicate very ordinary taste for you to use it. It is probably better to write on good, plain paper; although it could do no harm to have some such truthful phrase as the following printed on the back of each prescription blank, for the benefit of the public and the protection of your own interests: "A remedy that is useful for a patient at one time may be improper for the same patient at another time, or for other persons at any time, even though suffering with a similar affection."

Plain white-paper clippings suitable for prescription blanks can be purchased cheaply at any printing-office or book-bindery, or you can buy a ream of suitable paper from wholesale paper dealers, who will cut it into any size you wish.

It would be wrong, *very wrong*, to work hand-in-hand with a pharmacist, and receive from him a percentage on your prescriptions for sending them to his store, and for this reason: were you to accept part, it would be robbing either the pharmacist or the patient. Were the former to allow you so much for each prescription, and re-imburse himself by adding the extra amount to the sum charged the patient for the remedy, it could not be looked upon in any other light than that you had combined to *fleece* an extra amount from every unfortunate who trusted to your honor, just as one would look upon a lawyer who took fees from both sides. On the other hand, if the pharmacist possessed more honesty than you and allowed you to reduce his legitimate profit, because compelled to do so or lose your influence, it would place you in a most contemptible position, and you would live in constant danger of exposure and a public condemnation that the strength of Hercules could not, and the God of Justice would not, silence.

Honesty is the true keystone, without which the whole arch of honor falls.

"If I lose my honor I lose myself!"

You must live, and must have fees to enable you to do so, but, unless you obtain every dollar and every dime honestly and honorably, you cannot escape the finger of scorn; therefore, watch zealously that the public do not imbibe a belief that you are a part owner of or are interested in the loaves and fishes of the drug-store which compounds the largest number of your prescriptions. If such a suspicion be expressed by any one, or if any one insinuate that you seem to prescribe for the purse of the pharmacist rather than for the health of the patient, take care to inform him that you have no such interest.

If any pharmacist volunteer to supply a physician and his immediate family with medicines either free or at a nominal price, or with such proprietary or other articles as he needs, at cost, the favor can be conscientiously accepted, but it would be unjust to allow him to supply uncles, aunts, and cousins on similar terms. Bear in mind that such a course naturally entails more or less obligation or reciprocal professional attendance on the pharmacist and his family, and should be taken into consideration when accepting favors.

Duty, alike to yourself, your patients, and the profession, forbids you to supply one or several pharmacists with private marks, technical terms, or hieroglyphic symbols that other pharmacists cannot understand, as it would at once suggest trickery and corrupt motives. A still meaner (swindling) device would be to have a secret or cabalistic code, for use between physician and pharmacist, intelligible to them alone. Surely, neither you nor any other honest person needs warning against such abuses as these, for any one who would resort to private codes or cipher prescriptions for money-getting is neither honorable nor honest, and might very properly be classed with the vulture who rejoices at sickness, and the wretch who desires the epidemic. The trail of the serpent is over them all; knaves—

> "Whom none can love, whom none can thank,
> Creation's blot, creation's blank."

Your prescription is intended simply to tell the pharmacist what medicine you wish the patient to receive. When sent to the pharmacist it is an order for a certain medicine prepared in a certain way. The law has decided that this prescription or order belongs to the patient; the pharmacist, after compounding it, has, however, a natural right to retain it as his voucher, but he has no right to refill your order without your consent.

The unauthorized refilling of prescriptions by pharmacists has often produced the opium, alcohol, cocaine, chloral, and other enslaving habits. We also well know that it is often unsafe for a person to take a medicine ordered for another, or even the same medicine at different times. Furthermore, how can the pharmacist conscientiously label the second quantity, "Take as directed by Dr. Faraway," when Dr. Faraway is not even aware of the refilling?

In consequence of the present unfair habit of many pharmacists, the unauthorized refilled prescriptions probably outnumber those of the authorized, five to one.

Drug-stores have become so numerous of late, and the area from which each must derive its patronage and support is so limited, that their proprietors, in order to keep their heads above water, have either to charge very high for the medicines prescribed or *substitute* inferior drugs; the result is that drug-bills have gradually grown greater and greater, till of late they almost eclipse the charges for medical attendance. Many people, to avoid what appear to them *exorbitant* prices, now actually buy this, that, or the other quack medicine, make home mixtures, wend their way to no-drug irregulars or some over-the-counter-prescribing druggist, or trust entirely to nature, instead of paying physicians for prescriptions and then having to pay heavily to have them compounded.

The cost of medicines may be slightly reduced by instructing your patient to save the cost of the bottle by carrying one *with* the prescription; doing so cannot be objectionable to pharmacists, as they charge only *cost price* for bottles. A good and legitimate way to lessen the cost of certain prescriptions is to omit inert and unessential ingredients; for example, if you prescribe a mixture of wine of colchicum-root, tincture of digitalis, and sulphate of morphia for a patient, do not increase what would naturally be a one-ounce mixture, that would cost about thirty-five cents, into six or eight ounces, by adding syrup, water, or other vehicle, thus swelling the dose to a tablespoonful and the cost to a dollar. Prescribe the essential ingredients only, and let the directions specify how many drops to take and how and when.

A dose of medicine in powder or pill form is usually more expensive than the same in fluid form; besides, poisons and very active remedies can be more accurately divided when in solution.

Another evil resulting from there being too many pharmacists for all to live by legitimate business is, that not a few, not content with the great *"apothecaries' profit"* derived from the sale of medicines, encroach on the domain of medical practice, and prescribe, by the smattering of knowledge they pick up from the prescriptions of competent physicians, for every foolish applicant whose case does not appear to be formidable; even selling, by guess-work, this, that, or the other thing for *home cases* which they have not even seen, because asked to do so by the foolish; and thus build up a large office (or store) practice. How many, how very many, simple, functional cases are thus given medicines which do no good, but great harm, by taking the place of others which might have been of great benefit if given at the proper time, and are in this way, during the first few hours or days, converted into incurable or organic ones by such "medicine-men"; and how many new ailments are induced by Mr. Emetic's, Mr. Gargle's, and Mr. Jackall's haphazard prescribing heaven only knows. Fully one-half of all cases of venereal disease, biliousness, debility, cough, and the like, are now seen and treated by pharmacists (and their clerks and greenhorn apprentices) before calling on physicians. Four out of five of those whose complaints prove simple are, of course, cured like magic by the *four little pills* which the pharmacist recommends, or by the great liniment he sells, or by *his* noted fever-and-ague mixture or equally famous tonic, or his universal elixir, *that is simple and can't do any harm*, etc.; and they, thinking that he has turned some dire disease aside, laud the pharmacist to the skies and advise all to go—

"Fools go in throngs"—

to HIM for their livers, and kidneys, and lungs, and brains, and stomachs, instead of consulting a legitimate physician, with assurances that HE is as good as any doctor, and a great deal cheaper.

Hear Shelley in his scenes from the "Chalderon Dialogue":—

"*Cy.* Have you studied much?

"*De.* No: and yet I know enough not to be wholly ignorant.

"*Cy.* Pray, Sir, what sciences may you know?

"*De.* Many.

"*Cy.* Alas! much pains must we expend on *one* alone, and even then attain it not; but you have the presumption to assert that you know many without study.

"*De.* And with truth, for in the country whence I come sciences require no learning; they are known.

"*Cy.* Oh! would I were of that bright country! for in this the more we study we the more discover our great ignorance."

No person who is incompetent to examine a patient is competent to prescribe for him; and I would ask what sensible pharmacist would trust himself, or his wife, or his child to the examination and "subscriptions" of a neighboring pharmacist?

Another, although lesser, evil is this: If a patient's better sense carries him, in the first place, to a physician for advice, instead of to a pharmacist, ten to one he will be presented at the drug-store with one or two quack almanacs filled with infamous and alarming falsehoods, or a handful of advertising pictures, or that the bottle of medicine will be wrapped in Foolembad's or some other pushing fellow's handbill. The co-operation of the pharmacist as retailing agent for quack medicines is indispensable to quackery; and without it seven-eighths of the harm that patent-medicine literature is doing would cease, the vain promises that keep the public rushing from one lying wonder to another would no longer entice, and at least two-thirds of the quack and humbugging proprietary trash that now curses our land would slink from sight.

"Oh, where is the still, small voice of conscience?"

You will do well to avoid, as far as possible, all pharmacists whose presumption leads them to assume the rôle of a physician. The recommendation does *not*, of course, refer to *emergencies*, in which a pharmacist acts as a humanitarian. The manufacture of steel is one thing, and applying watch-springs is another. Medicines are the physician's two-edged tools; a pharmacist may prepare them and handle them for a life-time and be an excellent compounder, and yet, as his studies are pharmaceutical and not therapeutical, he may know no more about prescribing for the sick properly than the mechanic who makes needles or scissors does about dressmaking; or the instrument-maker does about operative surgery; or the manufacturer of trowels and plows and chisels about bricklaying, farming, or carpentering.

If a sick person ask a pharmacist for a plaster, a dose of cathartic pills, or an ounce of tincture of iron, there is no reason why he should refuse to sell them; but if he ask him what is the best remedy for this, that, or his other affliction, with a view to purchase whatever he designates, that is another, a *therapeutical* matter, and is beyond his sphere.

"Michael, Michael, you have no bees, and yet you sell honey."

Be also on your guard against instrument-makers and dealers who meddle with surgical cases, and manufacturers of appliances for deformities, examining or prescribing opticians, masseurs, etc., who presume to treat cases that should be referred to the physician or surgeon. In fact, avoid encouraging any one who encroaches on the physician's province.

Every patient should be warned that it is dangerous to wear spectacles, trusses, supporters, braces, pessaries, and the like, that have not been prescribed by a physician.

Make it a point never to style a pharmacist, an optician, a preacher, or any one else, " Doctor," or " Professor," unless he *be* one. Heaven knows the much-abused titles are cheap and promiscuous enough without bestowing them on ignorant spectacle-pedlers, and others who have not even applied for them.

Avoid overpraising any prescribing pharmacist to your patients, or people will, on your word, overestimate him, and begin to rely on his gratuitous advice, instead of on the physician's, in all cases considered moderate.

Beware of pharmacists who indiscreetly talk too freely, or converse, joke, etc., while compounding prescriptions, or knowingly insinuate to those who carry them prescriptions that they know what they are for, and have extra impudence when cubebs, ergot, etc., are ordered; or suggest to purchasers that the dose prescribed is too large or too small; also, the blundering blockheads who misread prescriptions or miscopy directions, or put wrong directions or the wrong physician's name on bottles, or surprise and alarm people by charging a different price every time a prescription is renewed, as if they had no system, or as if the medicines were put up wrong; who make the impression that it takes them half their time to correct the blunders and mistakes of the other half; who leave prescriptions partly compounded to wait on other customers, or to unscrew soda-water for sports who are in a hurry; or who in other ways allow interruption, or show abstraction or careless compounding. For such people be especially careful how you abbreviate, and how you make your ʒ's and 3's, and carefully dot every *i* and cross every *t* in your prescriptions, so as to afford them no shelter if a mistake occur, and, above all, to prevent a coroner's jury; or to clear yourself if a death-certificate is made necessary.

Mistakes in writing and in compounding prescriptions occur more often from improper haste, and by trying to do two or three things at once, than from incompetency.

Prescriptions written with ink instead of pencil have the decided advantage that they are not easily defaced and do not admit of easy erasure, etc.

A very good and safe rule in prescription-writing is to put down all the ingredients first; next write the directions to the pharmacist and the directions for use; then the number of doses should be decided on, and, lastly, the quantity of each ingredient should be carefully calculated and carefully written, followed by your name or initials.

Look on the back of every prescription paper you use to see that there is nothing of a mistake-causing nature accidentally written on it.

If you believe on good authority that any pharmacist so far forgets himself as to make disparaging comments upon you, or your professional ability, or your remedies, doses, or apparent inconsistencies; or to exhibit and decry your prescriptions to Irregulars, laymen, or other physicians, or to predict that they will not prove useful; or to make unauthorized substitutions, give under-weight of expensive ingredients, or omit them altogether,—

"Who knows the right, and yet the wrong pursues,"—

or to join with our enemies in reviling our profession and its imperfections, or in nick-naming different physicians in derision; or to keep his prescription-file open to miscellaneous inspection, or to have a medical *protégé* under his wing, into whose hands he endeavors to direct customers for selfish purposes, or to be guilty of any other grossly unprofessional conduct, you will be fully justified in directing your patients to go elsewhere for medicines.

In ordering syringes, brushes, atomizers, breast-pumps, probangs, etc., with your prescriptions, be careful to specify the kind or size you wish. To write a prescription for a solution, and add, "also a syringe for using," is often as perplexing to the pharmacist as if you were to send for a slip of adhesive plaster as long as a string or for a lump of rhubarb the size of a piece of chalk.

When any one is unable to pay the full price for what you prescribe, the words "Poor patient" in your handwriting, at the top of the prescription, will secure from any pharmacist the greatest reduction in price that he can afford to make.

You may take the following as somewhat of a guide in determining whether this or that pharmacy is conducted on a proper plane and worthy of confidence. Among the distinguishing features of a legitimate and properly conducted pharmacy are:—

1. Proprietor an experienced practical pharmacist, of intelligence, capacity, and integrity.

2. Competent and courteous assistants.

3. Pride and skill shown in selecting and preparing pure medicines.

4. Prescriptions compounded only by graduates in pharmacy.

5. A full and comprehensive line of pure drugs, apparatus and appliances for use in the care of the sick, also dietetics and sick-room conveniences kept.

6. An orderly and perfectly equipped prescription department. Store neat and attractive.

7. Quiet and discipline maintained. No loungers or smokers.

8. No liquors sold as beverages.

9. Not a bazaar of general merchandise.

10. Patent medicines and other nostrums shown and sold only when called for.

11. No habitual prescribing or giving medical advice.

12. Prices neither cheap nor exorbitant, but reasonable.

13. Prompt attention and accuracy characteristic.

Among the features that mark improperly conducted ones are:—

1. Habitual prescribing over the counter.

2. Indiscriminate refilling of prescriptions.

3. Unnecessary delay and detention of customers.

4. Careless handling of medicines and loose management of store.

5. Patent and proprietary remedies paraded and pushed.

6. Disparagement of physicians to the laity.

7. Store a resort for political or other crowds or cliques.

8. Unchaste conversations and disreputable conduct.

9. Wines and liquors sold as beverages.

10. Dealing in articles used for criminal or immoral purposes.

11. Engrossing attention to sale of soda-water, cigars, tobacco, fancy goods, etc.

12. Store kept merely as an adjunct to some other project.

13. Lack of sobriety in proprietor or clerks.

\* \* \*

Be prompt and decided in refusing to give laudatory professional certificates to any secret article; do not be too liberal even in giving them to legitimate pharmaceuticals, and never issue one founded on any other basis than purity of ingredients, or special skill or experience in compounding them.

Willingness to give medical certificates is an almost universal weakness of mankind. The idea of being paraded in print as "an authority" in connection with some wonderful cure is pleasing to thousands of people in every station of life, and makes them willing to have their names and even their pictures paraded in almanacs, hand-bills, and newspapers. Indeed, many impressible people, whose bump of wonder is easily touched, could almost be inveigled into certifying in medical matters that two and two make five by any sharper who understands how to tickle their self-conceit and love of notoriety.

Be alike determined in declining to give (un)professional certificates to any one on disputed or partisan questions, or in regard to surgical appliances, copyrighted medicines, rival wines, competing mineral waters, beef-extracts, baking powders, articles of commerce, patent contrivances, health resorts, etc., for they are often improperly used and made subservient to purposes not anticipated, and will affect the interests of the profession at large, as well as your own. If you ever give one, people who happen to know you may regard its personal and not its professional significance, but every one else throughout the land will know your title only. When amiable John Doe gives his certified opinion that ice is *hot* and fire is *cold*, it remains simply John Doe's opinion; but when John suffixes his title of M.D., he undoubtedly gives that certificate a professional significance, and, to some extent, involves the entire profession therein.

You may judge certificate-giving by its prejudicial effects on our own profession. One of the worst inflictions we endure to-day is the endless parade of misleading certificates from wide-mouthed clergymen, politicians, merchants, lawyers, D.D.s, LL.D.s (A.S.S.s, N.G.s), and other "distinguished citizens," known and unknown, recommending all kinds of medical nostrums.

"Heigh ho, the devil must be dead."

You know, and every sensible person knows, that such Pecksniffian certificates are not worthy of credence, and that the preacher of Gospel truth who (instead of confining himself to preaching the Glorious undefiled Gospel of the blessed God, the God of the Bible), bribed by a box of pills, or a bottle of bitters (that make drunkards and kill forty times as quick as whisky does), forgets his high mission, the cure of dying and perishing souls, and with reverential sanctimonious solemnity (ahem!) turns up the whites of his eyes,—

<blockquote>"O hollow, hollow, hollow!"—</blockquote>

and lends his name and the cloak of theology to assist the Diabology of charlatans and sharpers who deceive the afflicted with quack nostrums that are not worth the cost of the bottle they are in, must be either a silly dupe or a cruel knave.

<blockquote>"Knaves and fools divide the world."</blockquote>

Prof. Brass, Dr. Skinem, and every other sharp quack knows the influence of a clergyman's religio-medical indorsement published in a Sunday paper, and hence makes special and too often successful efforts to obtain it, feeling certain that they can easily entrap the dupable portion of the flock after the Shepherd (?) is secured,—

<blockquote>"He steers his boat well,"—</blockquote>

and it is a singular fact that, though few men get more gratuitous advice out of physicians than ministers of the Gospel, yet no class do more to injure the profession, by the ridiculous countenance they give to various kinds of quackery and pathies and isms. Truth should teach teachers to teach truth.

Suppose it suited the pride and the principles of our profession to enter the self-advertising arena, with quacks and patent-medicine men, and to scatter reports of all our daily cures and successes all over the land! Where would the petty triumphs of quackery, and patent pills, and bottled nostrums stand in the contest? Austin Flint *vs.* Hostetter; Samuel D. Gross *vs.* Brandeth, Johns Hopkins Hospital *vs.* Keeley's.

Whenever you are asked by traders in medicine, or their plausible drummers, who have no further interest in sickness than as it advances the sale of their nostrums; and when tempted by glowing advertisements, highly-colored certificates, epitomized treatises on therapeutics and practice, etc., to prescribe and make a market for their semi-secret trade-mark pharmaceuticals, copyrighted medicines, and the nine hundred and ninety-nine elixirs, restoratives, tonics, panaceas, and other specialties with attractive empirical names, gotten up by middlemen, crusading druggists, manufacturing pharmacists, and pharmaceutical associations, with labels that give suggestions for their use, to catch the popular eye and the popular dollar—think of the cunning cuckoo (see p. 32), and how its one egg hatches evil to the whole nest, and do not use them. Patent medicines are wolves in wolves' clothing; proprietary medicines are wolves in sheep's clothing, whose owners are begging favors from you with one hand and intercepting your patients with the other.

To fully realize the colossal proportions of the lucrative proprietary remedy method of superseding physicians, and of the mercenary motives and humbuggery that lie at the bottom of it, and the injury it inflicts on health, credit, and business, go and take a bird's-eye view of the vast and bewildering array of empirical and proprietary compounds: syrups, balsams, expectorants, and panaceas, each good for everything,—asthma and sore eyes, the itch and worms; and at the bushels of recommendations under which the shelves in the quack and proprietary departments of every wholesale drug-store groan, and then reflect on the enormous sums of money spent in telling—

Quack! Quack!! Quack!!!—

of their virtues in the newspapers, and on rocks, fences, and dead walls. Thus enlightened, you can hardly fail firmly to resolve henceforth to abjure them.

"The path of duty is the path of safety."

Unless you have mistaken your profession, are incapable of thinking and lack ingenuity, our standard and accepted agents, the United States Pharmacopœia and the dispensatories, should certainly be large enough and reliable enough to allow you to exercise yourself freely in the art of prescribing, and to make *any* required combination, and to accurately adjust the relative proportion of every ingredient to the condition of your patient; and you should, therefore, assert your intelligence and follow this, the legitimate mode of prescribing, and let our commercial rival's ready-made novelties, patented articles, and dishwater substitutes for medical attendance alone.

Of course, if anything truly useful or unmistakably better than the old is discovered, but not yet in the pharmacopœia, you would not, you should not fail at once to give your patient the benefit of it; but beware of all articles that are being advertised and pushed on catchpenny principles.

The principle which governs our condemnation of secret nostrums is this: They not only do more harm than good, but, if puffing and advertising alone enable the proprietor of a quack remedy to fleece the sick, its unprincipled owner deserves exposure and contempt. If the nostrum is really valuable, *which is very rarely the case*, its composition should be freely and fully disclosed for the benefit of suffering humanity.

You should also maintain your independence and never order A.'s, B.'s, or C.'s make of anything *unless* you have some specific therapeutical reason for so doing. To thus particularize would not only reflect injuriously on every other manufacturer and cause a still greater popular distrust of our materia medica and pharmacopœia, but also put the compounder to additional trouble and expense; for he might have several other varieties of the same article in his stock, and yet be compelled by your specification to get another. It almost invites substitution. I knew one case in which the pharmacist, though he had twenty-one different preparations of codliver-oil emulsions, resembling each other so closely in all important respects that but a hair divided them, standing spoiling on his shelves, had to get the twenty-second to fill such a prescription.

Do not, however, oppose any remedial agent that is a

distinct improvement in pharmacy, or any particular brand of anything on account of its being a monopoly, if that monopoly is owing to unusual skill, superior quality of materials used, or great perfection in its manufacture.

Patients are under the impression that pharmacists have about ninety cents profit in every dollar, and also think physicians know precisely what a medicine ought to cost, and will often ask you *how much* the druggist will charge for the remedies you have prescribed. Reply promptly that you do not know, that some medicines cost the pharmacist twenty times as much as others, and avoid mentioning any specific sum; because, were you to guess too high, they might infer that he had either made a mistake or used inferior drugs; and were you to guess too low, they would probably accuse the pharmacist of overcharging, and perhaps drag your name into their squabbles. Further, the people naturally overlook one all-important, priceless ingredient that every good pharmacist employs in compounding prescriptions, the worth of which he justly adds: I mean, the concentrated extract of brains.

Whenever you prescribe a remedy that is unusually expensive, such as musk, salicin, resorcin, salol, oil of erigeron, etc., take care to inform the patient of the fact, and that expensive drugs are no more profitable to the pharmacist than cheaper ones, so that he will not be surprised and cavil when the pharmacist tells him how much he charges for it.

Notice particularly whether a pharmacist gives unusual prominence to nostrums, quack almanacs and placards, or has quack advertising signs painted on his doors or outside walls, and it will give you a true insight into his aims and attitude toward our profession. If you see that he is pushing his *quack* department in a hurrah way, with quack proprietors' portraits in his windows and hanging around his store,—

"Roaring, roaring, roaring, nothing but roaring,"—

and his own name and influence used in handbills and almanacs as a vendor of nostrums, bitters, plasters, pads, etc., or selling liquor as a beverage, or selling medicines at retail or less than his pharmaceutical neighbors pay for them at wholesale, you may be sure that he is conducting his establishment simply as a tradesman, on a *trade basis* rather than on a professional one, which latter presumes him to love pharmacy and to devote his chief attention to the inspection and preparation of pure and reliable drugs, and compounding prescriptions with scrupulous exactness; and by shunning him you will fulfill a moral obligation.

To sell abortifacients, or vile nostrums intended to produce abortion, with the pretended caution, "Perfectly harmless, but not to be taken by women in a certain condition," is criminal.

"Cunning has but little honor."

Probably you have no right to ask or expect that the pharmacist should not deal in quack or proprietary medicines, or anything else for which there is a demand, as he keeps his store to make a living; you have, however, an undoubted right to expect him to show the equity of his position between their owners and us by keeping them out of sight, to be shown only when called for, just as he does sweet spirits of nitre, syrup of the iodide of iron, aromatic spirits of ammonia, and other fruits of pharmaceutical chemistry, instead of pushing their sale by displaying their announcements far more prominently than legitimate pharmaceuticals.

In drugs and medicines purity and accuracy are of the first importance, because the uniformity in action of every medicine is in proportion to its purity and goodness; some of our important remedies vary greatly in quality and in strength, and this is one of the occasional causes of uncertainty in the practice of medicine, and such variability would modify your efforts too much to be risked in any important case. A badly compounded prescription may rob you of your reputation and deprive the patient of his chances of recovery. If you think, therefore, that an important prescription is likely to be sent to a pharmacist whom you conscientiously believe to use inferior, stale, or impure articles, it is your duty to take care that it be sent elsewhere;

for, being responsible for the patient's welfare, and having your own reputation to care for, you have a perfect right, and indeed it is your duty under such circumstances, to have your remedies procured where you believe your prescriptions will be properly made up.

Pharmacy requires nice and delicate skill and imposes great responsibility, and the art of medicine is imperfect enough at best, and you will encounter more than enough of new and strange problems to remind you of your lack of aids and of the insufficiency of human resources, without adding the risk of being thwarted by the error, fraud, or accident of an unreliable pharmacist with deteriorated, adulterated, or inert drugs; but when you find it necessary to *ignore* any one for this reason, take care to do so in a discreet, ethical manner, with as little personality as possible.

Whether to allow a patient to know the name and nature and action of the remedies you prescribe, or not, requires great discretion, and good judgment is required to distinguish between persons who would and those who would not be benefited by an explanation of the intended remedies. There is often a temptation to endeavor to enlist the patient's confidence by furnishing him an insight into the nature and object of the agents employed; but the majority of experienced physicians seldom commit themselves, or if, in certain cases, to gratify the patient's whims, they appear to yield to the temptation, their explanations are advisedly ambiguous, and you, while judiciously seeking to inspire confidence in your patients, had better keep them, as far as may be, in ignorance of the remedies employed. But few physicians have escaped the chagrin of seeing their reasons and their remedies made use of to blame them and to cast discredit on their skill. You will, indeed, often wish you had synonyms for the terms quinia, zinc, opium, chloral, strychnia, morphia, and probably for other articles in daily use. Whenever a synonym for any of them is supplied, it will be judicious in many cases to use it. By employing the terms ac. phenic.

for carbolic acid, secale cornut. for ergot, kalium for potassium, natrum for sodium, chinin for quinia, tinctura thebaica for tinctura opii, etc., you will debar many a patient from reading your prescriptions and hampering you,—a check which is often highly desirable. You can also further eclipse his wisdom by transposing the terms you use from the usual order and writing the adjective in full and abbreviating the noun,—*e.g.*, instead of writing quiniæ sulph., write sulphatis quin.; compound cathartic pills, cath. pil. comp., etc., etc.

The official pharmacopœia distinctly recognizes the necessity of concealing the nature of certain preparations; and opium may be ordered under several synonyms without giving the slightest suspicion of its presence. You cannot greatly err in honestly seeking to conceal from your patients the nature of the remedies prescribed for their ailments.

"The silent physician has many advantages."

Be very careful to have all powerful remedies intended for external use labeled "For external use," or "Not to be taken," which will not only tend to prevent errors and misunderstandings, but in case they are swallowed by mistake it will save you from censure. For the same reasons, also be careful to order all mixtures, that may separate on standing, to be shaken before pouring out the dose, otherwise the patient may get all the active ingredients in either the first few or the last few doses.

When you prescribe a remedy of such an active character that it would poison if taken in large doses, or all at once, it is wise to make such verbal cautionary remarks about it as will fully put those who administer it on their guard. Also, when you prescribe a remedy for external use, and at the same time one that is to be taken internally, be careful to tell the patient how each will look and smell, so that he may not confound them and swallow the wrong one. Absent-minded pharmacists have more than once put liniment labels on bottles containing remedies for internal use, and those designed for the latter upon the liniment-bottles, thereby leading to a jury of inquest, which a word of explanation from the physician to the patient might have prevented.

Pharmacists might easily avoid the possibility of thus exchanging labels by compounding one and labeling it before commencing the other. By instructing the pharmacist to put a *red* label on all the bottles for external use, security against mistakes is better insured.

If, in prescribing such agents as tincture of belladonna or tincture of iodine, for external use, you direct the pharmacist to "put brush in the cork," seeing the brush when the bottle is opened, will almost surely prevent its being taken internally.

You will notice that some pharmacists label the remedies they compound for you with their *file numbers only*, thus, 7483; while others adopt the much more satisfactory plan of adding the date on which it was compounded, thus, 7483, 19–7–93, signifying that it is numbered 7483, and that it was compounded July 19, 1893. The latter plan will enable you to distinguish between the dates at which you prescribed different bottles of medicine, and may otherwise be of service to you. I am quite sure the majority of pharmacists would cheerfully make use of this system if they were aware how often it assists the physician.

Even with the best care every one is liable to make mistakes, and even the wisest men are not always wise. One might write tablespoonful where he meant teaspoonful, or sulph. morph. instead of sulph. quin., or acid. carbolic. when he meant acid. boracic., or tinct. opii when he meant tinct. opii camph., etc.

It is well, therefore, to request neighboring pharmacists always to inform you of any ambiguity or apparent mistake in prescriptions bearing your initials before dispensing them, and, in return, when

"Some one has blundered,"

and you have reason to suspect the mistake has been in compounding the prescription, be careful not to make your suspicion known either by word, look, or action, till you have conferred with the person who dispensed it. The error, if one exist, is just as apt to be yours as his.

When a prescription is for an infant, or a young child, it is a great safeguard against error in compounding to put at the head of the prescription, "For an infant," or "For a child," or "For little Willie," etc.

Bear in mind that the pharmacist, like yourself, is only human, with long hours and short pay, and that he, like other persons, requires some *rest* and relaxation from his drug-mixing and drug-selling slavery; and do not order mixtures requiring tedious manipulations, or direct filthy ointments to be made, or dirty plasters to be spread, suppositories to be molded, or other unpleasant duties to be performed on Sunday, or during sleeping-hours, unless they be urgently needed.

# CHAPTER XI.

*"Sound policy is never at variance with substantial justice."*

As a physician you will hold two positions in relation to patients: first, during sickness you will feel a humane interest in them and a scientific interest in their diseases, give them your best skill and your labor, and employ whatever remedies will be most surely, most safely, and most rapidly beneficial; to this you will add sincere sympathy and commiseration. Later, when, by recovery or death, your interest and skill are no longer required, you will enter upon the second, or business relation, and then you should, unless poverty forbid, demand and secure, in a business-like manner, a just remuneration for your services.

Business is business, and should always be regarded as such. You must be clothed and fed, and must support those dependent upon you, just as other people do. Every person naturally and properly looks to whatever occupation he follows for support; therefore, let not false delicacy or out-of-place politeness break up the business part of your profession, or interfere with your rules in money matters, or prevent your knowing where sentiment ends and business begins. You are human, and must live by your practice, just as the priest lives by the altar, the lawyer by the bar, and all other people by their avocations. The practice of medicine is the work of your life; it is as honest, useful, and legitimate a branch of human industry as any other on the face of the globe, and no one earns his means of living more fairly, and often more dearly, than the hard-worked physician, and both common sense and vital necessity require that you should try to provide properly for yourself and for those dependent on your labors for support.

This you cannot do unless you have a business system, for upon *system* depends both your professional and your financial success. No man is at his best when handicapped by poverty; and no one can practice medicine with clearness and penetration, earnestness and effect, if his mind be depressed and distracted, or health lowered and temper vexed by the debts he owes, or be annoyed and dunned by hungry creditors at every corner; or whose discontented stomach is uncertain where the next meal, for himself and his care-worn family, is to come from; or who walks the floor and knows not which knock at the door will be the sheriff's. These and other cares, that poverty entails, dwarf any (Deadbroke) physician's mind and body, and cripple his work; and it is only when free from the incubus, the mental solicitude of debt and poverty, that his mind and his energies can do full justice to his attainments.

"Anticipated rents and bills unpaid
Force many a doctor into the shade."

In these days neither untiring study, nor unselfish devotion as a humanitarian, nor the bubble of applause will enable you to live on wind,—

"All leaf and no fruit,"—

or lift you above the demands of the tailor, the instrument-maker, the book-seller, the grocer, the butcher, and other creditors, not one of whom would accept your reputation for professional devotion, or of working for philanthropy, or your smiles, thanks, and blessings for his pay; nay, even the conductor will repudiate such sentimental notions, and put you off the street-car which is carrying you to your patient, if you do not have money to pay your fare.

"Wrinkled purses make wrinkled faces."

It is, naturally, a pleasant thing to be *very popular;* but even were your air-built popularity and verbal fame to embrace the whole city, neither it nor checks on the Bank of Fame will fill your market-basket nor purchase books, pay your rent nor feed your horse; and although the Glittering Dust is neither the primary nor the chief incentive in the practice of medicine, it ever has been and ever must be one of the objects, for no one can sustain his practice without a money feature.

"Necessity has sharp teeth."

If people do not pay you you cannot live by your calling, and you will very soon tire of *all work and no pay*. Almost as well to starve without a patient.

In your money affairs be systematic and correct, for it is as important to charge your visits as it is to make them; make it your habit never to retire to bed without making some kind of record of every visit, etc., made during the day. The nearer your financial arrangements approach the *cash* system, the better it will be for you and your family. Frequent accounts are best for the physician. If he render bills promptly, it teaches people to look for them, and to prepare to pay them, just as promptly as they do other family expenses. It is often more advisable even to submit to a reduction in a bill for prompt payment, than to let the account stand over and run the risk of losing it through the pay-when-you-please system, for while you are waiting some may fail and others abscond. Besides, after settling promptly, many patients will feel free to send for you again and make another bill, even in moderate sickness, instead of dallying with home remedies or quack medicines, as they might do if they still owed you.

You should render your bills while they are small, and your services are still vividly remembered, not only because gratitude is the most evanescent of human emotion, but for another reason: if you are neglectful or shamefaced and do not send your bills promptly, it will create a belief that you do not believe in prompt collecting, or are not dependent upon your practice for a living, or have no wants and do not need money; or that you do not hold this or that person to your business rule, or are not uneasy about what *they* owe you; and if you foster a bad system of book-keeping a bad system of collecting will grow up around you, and great loss will result. Asking for payment reminds patients that there is still a little of the human left in a man, even if he have become a physician, and that, since you have to live, you must have your fees to enable you to do so.

The business of the world is now conducted on the *cash* system, instead of the old *long credit* plan, and you should do your share to

"Break the legs of the evil custom,"—

the unjust habit that physicians used to follow, either through carelessness or to maintain the favor of patients, of waiting six months or a year after rendering services before sending a bill. If a physician attend a person, say in February, and send his bill in March or April, it seems to the patient like a current expense, and as though the physician lives by his practice, and it is apt to be paid promptly; whereas, if he delay sending it until July or January, and then send one headed with the semi-apology, "Bills rendered January 1st and July 1st," as an excuse for even sending it then, the debtor will naturally think that the physician has merely sent his out with a whole batch of others, *more* because he has posted his books than from a special desire for its payment; and in this belief he will probably let it remain unpaid for months longer, and perhaps delay its settlement till it becomes an old back debt, which is the hardest kind to pay. All sorts of strange accidents are continually happening that may prevent payment; besides, time effaces details, and recollection of the number of visits, the physician's watchings, cares, and anxieties are forgotten, responsive sensibility is lost, and the bill, though really moderate, is apt to look large. All these circumstances combined are apt to make people feel, when they do pay an old bill, not as though they are paying a well-earned fee, but more as if they are doing a generous thing and making the physician *a present* of that amount.

If, in spite of these facts, you do send your bills only every six months, instead of putting on them "Bills rendered every six months," put "All bills collected at the end of every six months," or "Prompt settlement of bills is kindly requested."

Also, show that you keep records of your cases and of your fees by having on all your bills the word FOLIO............ after the patient's name, with the number of his page inserted in the blank.

You will have to make considerable reduction in many large bills after they have become old; therefore, look after them while they are small in amount and recent in date. Indeed, if you let one bill be added to another till the total reaches a considerable amount, you may place it wholly beyond the power of the person to pay it, and wrongfully *force* him into the position of a dishonest man. Besides, long-standing bills frequently lead to a disruption of friendly feeling and loss of practice.

"Old reckonings breed new disputes."

The very best time to talk business, and have an understanding about fees with doubtful or strange patients, is at your first visit or the first office interview, and the best of all times to judge a person's true character will be not on occasions for social intercourse and the ordinary amenities of life, but when you touch his financial pocket-nerve and have money dealings with him.

"Then you will find out what stuff they're made of."

Even a single dollar will sometimes show you exactly what a person is, whether a knave or a man of honor.

Make it an invariable rule never to accept a commission or fee from any one under circumstances which you would not *willingly* submit to public exposure or investigation by a medical society, or a court of justice. Probably your severest test will be when money is enticingly offered to induce you to do doubtful things.

Many and many a patient will quit employing you to escape from paying an old bill, and then, to hide from their surprised neighbors the true cause of their quitting you, will trump up some falsehood or another, and give you a bad name, to prevent them from employing you and thereby possibly learning from your lips the true reason why they changed.

Railroad and steam-boat companies and other corporations, also proprietors of mills, factories, workshops, etc., whose employés get injured, in order to relieve themselves from responsibility, or from fear of incurring public odium, or from a selfish fear that they may become involved in suits for damages, and be made pecuniarily responsible for the injury, often send, directly or indirectly, for a physician to attend, and in one way or another create an impression in his mind that they will pay the bill, but afterward, on one plea or another (usually this—that they have supported the injured person during his disability, which is as much as they can afford), either entirely disclaim the debt or refuse to pay it, and with such excuses leave the physician in the lurch.

"Rank injustice that smells to heaven."

In such cases you may obviate this result and secure justice, or, at least, ascertain the prospect, by going, as soon as possible after you have taken charge and given the initial attention, directly to headquarters, or to the person who has the authority to make the company or firm financially responsible for your services, and, after explaining the labor and responsibility which the case involves, make known your doubts of not being recompensed for your services unless they will see to it, and frankly ask if they will assume the responsibility and let you enter the account on your books in *their* name.

From similar motives, the heads of families, for their own satisfaction, for social reasons, or from a feeling of insecurity lest some inmate of their house who has become sick may have a contagious disease, will sometimes request you to visit their servants, nurses, or maybe poor relatives, and then seek to avoid payment of your bill on one pretext or another. If there be reasonable doubt of prospective payment in these cases, you had better at once seek to determine the financial responsibility, as suggested in the preceding paragraph.

Bear in mind the fact that when a person, even though a banker or a millionaire, comes for you, or summons you, or requests you to attend another person, he is not thereby made legally responsible for your fees, unless he distinctly promises or agrees to be responsible for the debt. Hence, make it a rule to enter the names of those who are held financially responsible for such services in your book, and keep a memorandum of the facts that make them so, and make out your bill to them accordantly. If you take these precautions it will prevent many unpleasant misunderstandings, and save you many a hard-earned dollar.

Before you have practiced long you will find that your welfare will depend not upon how much you book, but upon how much you collect, and that if you never insist upon the payment of your fees you can never separate the wheat from the chaff. If you have a business rule, and people know it, they will associate you and your rule together, and be guided thereby. Let the public know, in the early years of your practice, what your rule or system is, or you cannot do so later in life. When a new family employs you, render your bill as soon after the services as the ordinary courtesies of life will allow, and especially if there have been a previous attendant who was a careless or indifferent collector, or no collector at all. Send it in as a test, and if there be any objection to you consequent on the early presentation of your bill, or because you want your fee, the sooner you arrive at an understanding of each other, or part company, the better for you.

Some physicians have more tact in getting fees than others, and, curiously enough, there are patients who will pay one physician but will not pay another, there being certain persons with whom they desire to stand well, and others for whose opinions they do not care. Try to be in the former class with all persons of doubtful integrity.

When patients ask you how much their bills are, or how much they are indebted for office consultations, operations, etc., always reply, with courteous promptness and decision, "one dollar," or "ten dollars," or whatever else the amount may be, large or small; and if you be careful to avoid prefacing or following this reply with other words, most people will, in the embarrassment of the moment, proceed to pay you without objection, whereas if you add more words it will weaken your claim in their minds, or impress them with the belief that you have no settled charge, and will furnish them with a pretext to show surprise and contend for a reduction. When one does demur at the amount, show your amazement, and be prepared at once to defend or explain the justice of the charge.

Your accounts for surgical cases, midwifery, poisoning cases, and, in fact, for all exceptional cases, should be promptly posted and charged in your ledger; otherwise, the patient may call unexpectedly to pay his bill, and you may, either through haste, or embarrassment, or temporary forgetfulness of all attendant circumstances, name much too low a figure and do yourself provoking injustice. Besides, the amount being already determined on and entered in your book shows it to be the settled charge, and the patient is less apt to ask for a reduction.

Take your fees for honest services whenever tendered. Patients will often ask, "Doctor, when shall I pay you?" or "Shall I pay you now?" A good plan is to answer promptly, "Well, I take money whenever I can get it; if you have it, you may pay it now, as it will leave no bones to pick," or "Short payments make long friends," or "Prompt pay is double pay, and causes the physician to think more of his patient," or something to that effect. Never give such answers as "Oh, any time will do!" or "It makes no difference when," or you will soon find it to be very expensive modesty.

Although Sunday is a holy day, on which bills should not be sent, yet it is perfectly right for physicians to accept fees earned or incidentally tendered on that day.

Never neglect regularly to post your account-books, for it would be violating nature's first law—which says that the first object of every being is to supply his own wants—to attend faithfully to the department of your occupation that concerns others and neglect the one that concerns yourself. The Scripture command is, " Love your neighbor as yourself ;" it does not say love him *more*, but Paul does say to Timothy: If any one provide not for his own, and specially those of his own house, he is worse than an infidel.

It is a good plan to insert the names of transient patients in your cash-book, instead of blurring your ledger with them, and to give pages in the latter only to probable permanent patients.

Try to get cash from strangers for catheterization, certificates, vaccination, and other minor services, instead of blurring your ledger with petty accounts.

When a prompt-paying patient pays cash at each visit, or settles at your last visit, so as to make it unnecessary to transfer his account from your visiting list to your ledger, the simplest way to mark it paid is to turn each visit-mark (*l*) on your book into a **P**, signifying *paid*.

A good plan to use in making out the list of calls you are to make each day, and the order in which you wish to make them, is this: Tear up a lot of foolscap or note-paper into slips as long as the page and half as wide, and draw a line down the middle of one side of each ; go over your list each morning, and cull out the names of all who are to be visited, and put them on one of these strips, left side of the line. Then select and arrange them carefully on the other side of the line in the exact order you wish to observe in visiting them, putting urgent cases and early calls at the top. Cut off this list when completed and carry it in your outer coat- or vest- pocket, refer to it often, and tear off each name as the visit is made.

You can readily fix your visiting-list so that it will always open at the page in use. To do this, clip off about half an inch of the upper corner of its front cover, thus ◥, and then in like manner cut off the corners of the leaves thereby exposed, down to the page corresponding to the date thereof. When thus prepared, if, in opening the book, you place your right thumb on the exposed corner of the uncut leaves, it must open at the proper page. As weeks pass, clip each page as required. The most convenient way to carry your visiting-list is in a wide but shallow pocket on the left hip.

Do no unnecessary bookkeeping, but take care to do enough to keep your accounts correctly. The visits and cash entries in your visiting-list and day-book should be written in ink; for, being original entries, they would be accepted in court as legal evidence. A good way to prevent any one or any thing being forgotten is to write names, visits, street promises, etc., in your visiting-list with a lead-pencil without delay, till you have a chance to rewrite them with ink.

Purple, green, and blue inks all fade badly, and occasion a great deal of trouble. You had better keep your books with good black ink.

At the end of every week add up the visits made to each patient whom you have attended during the week, and after ascertaining the total sum which you should charge therefor, insert that amount in the blank spaces found at the end of the lines after the Saturday column in the visiting-list. By doing this weekly you can fairly estimate and charge the value of your services to each patient while they are still fresh in your mind. It is not only wise to enter at the end of each week the amounts charged, but also to enter the names of the individual members of the family who have been under your care during the week, in the visiting-list over the visits, *for reference*, in case your attendance should ever be disputed.

In posting your books at the end of each month, in order to avoid missing any entry in transferring the items from your visiting-list to the ledger, make use of the simple checking-off plan. A good way is to make a list of the names of all patients whom you have treated during the month on a sheet of foolscap paper, then bring from the visiting-list to the foolscap the amounts marked against them for each week's services and put those of each after his name; when you have all the charges transferred in this way to the foolscap, begin and go over your ledger, page after page, and scan every account as you go along. When you reach the name of any one against whom you have a charge to make, add up all you have marked on the foolscap against him, and enter the total on his page of the ledger; but instead of wasting time to write November, 1892, $7.00, enter 11-92, $7.00, then cross that person's name off the foolscap list, and continue on, page after page, through the entire ledger. By this crossing-off system, if you chance to pass over any one's account, it will remain *uncrossed* when you are through the list, and will thus be detected. While going over the different pages of the ledger note down on the blank after the word folio, on one of the small pile of blank bills lying at hand for the purpose, the number of each one's page whose account needs RENDERING, so that on completing your entries you may readily return and make out the bills in question; also, take care while turning the pages to make a list of the indebted patients whose accounts it would be well for you or your collector to look after during the ensuing month.

When you make out a bill, enter in your ledger, in the space just after the amount, *the date* on which the bill for that amount was rendered; thus, $7.00, with 1-8-92 after it, would signify that a bill for seven dollars was rendered to that person on the first day of the eighth month, 1892; or it may be written as the Quakers do, month first, then day, and then year, thus: 8-1-92. Payments may be similarly entered.

A good way to save the trouble of looking over worthless or lapsed accounts in your ledger, month after month and year after year, is to cross them off, using lead-pencil, which can be erased at any time, if necessary, for such as may possibly be revived; and for those that are dead or, from other causes, never likely to employ you again, use ink.

That a patient whose name is on your books is a colored person can easily be indicated by putting three dots after his name, thus: Robinson, John, ⋮ 13 Columbia Street.

Patients will occasionally dispute the correctness or justness of your charges. If a bill be not correct, correct it at once and willingly, with such an expression of regret at the error as may be judicious; if, however, it be correct and just, do not allow yourself to be browbeaten into the position that it is otherwise. Many people are not aware that the charges for *surgical* and various other cases are higher than for ordinary visits; some appear to think that for a visit at which you reduce a dislocation, open a large abscess, make a vaginal examination, or draw off the urine, you should charge the same as for ordinary visits; others have an idea that physicians do not, or should not, charge for every visit when they make more than one visit in a day, or for every patient when more than one in a house is sick. You must, of course, correct their error by explaining the relative difference, or, if necessary, by reference to the fee-table.

Never undercharge for your services with a view of obtaining business, or in any other odious sense. A community never values a physician higher than he values himself; besides, habitual deviation from the uniform rate of charging is considered dishonorable and is ruinous to one's interests and to the interests of the profession at large. Moreover, the public knows that no man will be content with small and insufficient fees while his brethren are receiving greater, unless he rates his abilities at a less price. Small fees are, therefore, set off against small skill in the public belief. The tendency of undercharging is to put a lower value on the medical profession, to lower the fee-table permanently, and to compel all physicians to work for inadequate pay. There is a vast difference between underbidding in our profession and that seen in wars of competition in ordinary business pursuits. In the latter, underselling, cut-rates, and other results of severe and crushing competition are only temporary; for, if merchants or traders were to sell goods at or below cost for a length of time, failure would result. In commercial or business wars one or other withdraws, or they enter into a compromise and each advances again to full prices; snapping and snarling physicians, on the contrary, having no goods to manufacture or sell, one determined to triumph and the other resolved to prevail over his "opponent" (!) by underbidding and exposing each other's misfortunes, may keep up the strain of rivalry and efforts to crush or banish each other for years, dispensing their skill to everybody for insignificant or nominal fees, impoverishing one another, and almost starving those depending on them for support.

Besides,
"Wars bring scars."
"What can war, but endless war still breed?"

Surely we suffer enough annoyance in the proper pursuit of our profession, without adding to our troubles by such struggles.

Unless you already have a regular scale of charges in your region, try to bring about a somewhat uniform fee-table or rate of charging among the body of physicians.

The wisest rule in charging for your services is to do your work well, then ask, even from the beginning of your career, the fees usual for conscientious, skilled attendance,—neither exorbitantly high, like an extortioner, nor absurdly low. And always maintain that your services are as good as the best.

Let people know that you honestly strive to make your bills as small as possible, not by undercharging, but by getting them well by good treatment and with as few visits as possible.

Never enter into an auction bargain to attend a patient or a family by the week, month, or year; it is far better to be paid for what you actually do, than to have some people feel that they are giving you twenty dollars for five dollars' worth of service, while you, on the other hand, are, in many other exacting cases, giving fifty or a hundred dollars' worth of service for twenty dollars, and have no alternative but to fulfill the contract.

Also, never bargain to attend whole neighborhoods or clubs of poor people at reduced rates, or at half- or quarter- price, because your antiquated or unripe neighbor does; it is bad policy, and never works successfully. Indeed, if you ever attend a confinement or other case in a family for a nominal fee, or lump your bill for ready money, they will always expect to pay what they paid before, and you will not be able to raise your scale of charges to the regular price in that family after your standing and skill improve and your time becomes more valuable; or even with other patients who hear of it.

It is a mistake to think that you can greatly augment the charges you make in the beginning of your practice as you advance in age, skill, and experience, as everybody will appeal to your former charges, and object. After becoming accustomed to small prices, old patients will even think you ought to charge them less instead of more; so that, if ever you feel unwilling to repeat services of any kind for the sum received for a previous case, be careful to give the patient fair notice of your intention to raise your charges.

One of the hardships of our profession is that the older men, perhaps now rich, or deriving their support chiefly from their stocks, bonds, four-per-cents, or farms, continue to charge the low prices of half a century ago, while the price of living, etc., have all advanced; so that the younger physician, without these, must charge somewhat the same, and thus hardly get revenue enough to support him.

A wise man usually accommodates himself to circumstances and takes what he can get, but even when you are sure that, to meet one's means of remuneration, you will have to receipt your bill for a reduced amount, make it out for the standard amount, so that the debtor may see the real extent of his indebtedness and give you credit for the amount of the reduction; in other words, when you make a reduction to those who plead poverty or other acceptable reason, let them understand that you are not reducing your charges, but are taking something off their bill; and enjoin upon them not to tell it around, lest it lower your scale of charges elsewhere.

In attending an extraordinary case for Mr. Bullion, or Gov. Goldmine, or Gen. Doublebank, or Maj. Opulent, or Capt. Creamyrich, Mrs. Bountiful, or any one else who is very rich or notoriously liberal, after properly calling his attention to the immeasurable value of the life you have saved, or of the blessing and health your services have given, leave to him the money valuation of the benefits the restoration brings, or the worth of exemption from death, unless he insist on having a bill. In the latter case, charge him no more than any one else for the same services. In the former you may, by submitting it to him, from his feeling of superlative delight at the successful issue, be paid most munificently, possibly ten times as much as your bill would have been.

When people talk to you about taking off part of their bill because they are poor, and charging the rich more to make it up, take less if you think proper, but under no circumstances allow them to infer that you, or any other physician, would charge any one, whether rich or poor, a cent more than is honestly your due.

It is customary and just to charge a *double* fee for the first or for an only visit in a case, chiefly for the following reasons: You must at the first visit devote an extra amount of time and attention to learning the history of the case,—maybe make a minute time-consuming examination,—must involve yourself in a diagnosis, and probably also in a prognosis,—must carefully think over and decide upon a whole line of treatment,—must instruct the nurses,—map out the quality and quantity of diet, drink, exercise, etc.,—point out the requirements of hygiene, maybe institute asepsis or antisepsis,—lay down general rules regarding lighting, heating, and ventilation, the clothing, the temperature, the toilet, idiosyncrasies, etc., and formally establish yourself in the case, and assume all the responsibilities of the issue. These combined make it an extraordinary visit, and fully justify a double charge for the first visit.

The first visit to a case may be easily designated by turning the visit-mark (*I*) into an *F*.

It is also just to charge extra for a visit in which you are detained longer than (say) a half-hour, or in an obstetrical case over five or six hours, either by the urgency of the case or where the family request you to remain.

There are a few people who consider that when a case is serious enough to require the physician to make more than one visit a day he should not charge for the additional visits, unconscious, as it were, of the fact that cases dangerous enough to require an extra number of visits are the very ones which entail upon him the greatest responsibility, cause him most anxiety, and contribute most largely toward making his life one of wearying labor and self-denial.

When you attend two or more patients in a family at the same time, take care to charge full rates for one patient and half-rates for each of the others.

You will often have people to hum and haw, and complain that their bill is high, and ask you to make a reduction; yet, many of these very people would not employ you if you were a third-rate or low-priced physician. Everybody wants first-class services, but wants them as cheaply as possible. It is not human nature to prefer a fifty-cent to a two-dollar silk: but if people be lucky enough to get the two-dollar silk for one dollar, they congratulate themselves. They reason the same about physicians; very few prefer or appreciate a low-priced (cheap-John) physician.

In *unusually* severe cases, and in those which require very great exposure or *extraordinary* legal or professional responsibility: in cases of recovery after poisoning, or of apparent drowning, or suffocation, of small-pox and other loathsome and contagious diseases, the fear of which prevents other patients, who know you are attending them, from employing you, or which necessitate loss of time in changing your clothes and otherwise disinfecting yourself before visiting others who are not affected, or in which you have evinced remarkable skill, or where you have had very great luck in bad cases of any kind, you should charge good, round fees.

It is certainly worth far more successfully to attend an important or distinguished member of the community in a case of pneumonia—in which you save his life as clearly as if you had dragged him helpless from the flames, or plucked him drowning from the water; or a patient with apoplexy, or with a wound, ulcer, fracture, or a luxation, or a contagious disease in which you risk losing your own life; in fact, anything that causes you great anxiety and necessitates much study—than one for whom nobody cares, with a sore finger or toe, or chicken-pox, mumps, or hives, even though the two cases require an equal amount of time, or a like number of visits.

In some cases your charge will be not so much for the work actually performed as for your knowledge and skill in knowing how to do it; for instance, you may charge twenty dollars for the few minutes' work of reducing a luxated humerus; if this were duly itemized it might read thus: "For reducing dislocated shoulder, five dollars; for expense and study of learning how to do it, fifteen dollars." "You charge me fifty sequins," said a Venetian nobleman to a sculptor, "for a bust that cost you only ten days' labor." "You forget," replied the artist, "I have been thirty years learning how to make that bust in ten days.

Attendance on Bigbee's beloved child, on an eminent or very important member of the community, or on one of the great men of the land, for whose life you have fought a great battle, or on a well-satisfied stranger who has journeyed far with an important case that causes you special solicitude and anxiety, or on a case that presents peculiar difficulties, justifies you in making a special charge, whether attended at your own office or at the homes of the patients. In such cases pay every necessary attention, but be careful to make no unnecessary visits, unless by special request; for in a very important case, in which three visits would be really necessary, to which you make but three and then discharge yourself, your services will be appreciated more highly, and the family will more cheerfully pay a fee of a hundred dollars than if you had also made five additional, apparently unnecessary, visits, and charged but eighty dollars.

On the same principle, when you have severe cases of any kind that necessitate several visits in the course of the day, take care to diminish the number markedly as soon as the necessity ceases.

In extraordinary and complex cases; also, where the results are apt to be great and far-reaching, or in which you go a long distance, or at very unusual hours, or through great storms, or extra dangers, the charge should be not by the visit, but for the case.

Patients will often express surprise at your asking the same fee for office advice as for a visit to their house; explain to them that, although the charge is the same, it is much cheaper to be an office patient than to be visited at home, because an office patient usually comes but *once*, or *only* when his medicines are out, or when some important change has taken place in his ailment, and quits entirely as soon as possible; whereas, if you have him under care at home, your responsibility and feeling of uncertainty compel you to visit him frequently to ascertain whether he is getting along as expected. For these reasons a few office consultations with the responsibility of attending faithfully resting on the patient, if on either, often suffice, instead of many house visits, and in this way office advice becomes very much cheaper.

Some people who are mean and miser-like about paying —as big and exacting as tyrants when sick, and as small as potato-bugs at bill time—will want you to deduct largely from their bills, especially if they happen to be mostly for office consultations, vaccinations, and other services of a less important character. Meet them at once with the argument that if they are to pay you less than the average for the minor services, you will have to charge them on a much higher scale of fees for the more important ones. But with such people the question is not services, but money, and you will often have a stinted sum grudgingly given, even for the saving of life.

Be kind to the poor and lenient with the unfortunate, but when people are able you should be as rigid in requiring your pay as other men.

The difference between words used with your office patients will sometimes make all the difference between a fee and no fee. Some who consult you, if asked to call again *to let you know how they are getting on*, will, on returning, show by every word and action that they do not expect to pay, as they merely called because you requested them to do so. Therefore, unless you intend to omit the charge, it is better to *advise them to consult you again*, at such time as you deem proper to specify. This will distinctly intimate to them that your usual fee will be charged.

When a new patient, whose honesty you have reason to doubt, consults you at your office, and instead of paying the fee defers it, with a promise to call again, if you request his name and residence, and book them in his presence, your chances of getting paid will be greatly increased.

Never agree or enter into a contract to attend any one for a "contingent fee"; that is, do not take patients with chronic sores, constitutional headaches, epilepsy, cancer, post-nasal catarrh, pimpled faces, hæmorrhoids, dyspepsia, hypochondriasis, and other chronic affections; or victims of syphilis, gonorrhœa, the ruthless blight of scrofula, etc., on the "*no cure, no pay*" system, or to pay "*if their rainbow expectations are realized,*" or "*when all is over.*" Enter into no such one-sided agreements to do things that may prove impossible, for they are never satisfactory, and will generally end in your being swindled, and, it may be, charged with incompetence or malpractice. In expressing your willingness to undertake such case, let it be clearly understood that if the case be curable, then you are there to cure it, but that you *charge for services, not for results*, and must be paid for your attendance even though the patient proves incurable or dies, and that all who seek your advice *must* take the probabilities of cure or relief from your well-intended endeavors. Remember: having accepted charge of a case, you are morally bound, pay or no pay, conscientiously to fulfill your duty to the patient; you may, nevertheless, fairly intimate to those who you think are unworthy of credit, that if they pay as they go on, instead of running up a bill, it will tend to encourage and interest you more in the case, and naturally inspire and stimulate you to do your best.

Some persons suffering from constitutional syphilis, ulcerated legs, chronic eczema, broken constitution, etc., in which the treatment may extend through many months, or maybe for years, or even through life-time, will probably suggest that you should wait for your fees till done attending. Do no such foolish thing, as such a case may die, or move away, or abandon treatment, or slip away from you to another, or begin with grandmother remedies, or with " yarbs from those who have no larnin'," or even resist all your attempts to effect a cure, and you may get nothing except misrepresentation for all your work.

In such cases, it is far more just and wise to render your accounts at the proper time,—" for the three months ending ———," or, at the very furthest, the first day of every July and January. If they demur (which they cannot justly do) do not hesitate to express your surprise at their doing so, and, in reminding them of the necessity for living by your practice, cautiously but firmly tell them of your entire unwillingness or financial inability to allow your fees to accumulate as they suggest.

You should ordinarily exact no previous stipulation of pay, and manifest no undue anxiety in respect to your fees, and make no reference to your intended charges, unless you are dealing with people notoriously unworthy of confidence, or when a misunderstanding is apprehended; but in most instances, unless the patient be well known to you, you should not hesitate to require your fee *in advance* (your chance of compensation will grow worse as the patient grows better) for attending cases of *secret* diseases. If you fail to do so, Mr. Hightone or Mr. Lowtone, or Mr. Notoneatall, as the case may be, will almost certainly leave you, about the time that Richard's himself again, with his bill unpaid; and if you press him about it, he will either pay it grudgingly or not at all; and, should you dun him for it, will abuse you, and, with vinegar or ice in his looks, meanly assert that he is absolutely a Joseph, and that it was not an ignoble disease at all, but only a strain, or that you did him no good, or almost killed him; or tell some other falsehood as an excuse for deserting and trying to defraud you, and ever after try to bring you into public odium and to injure you to the extent of his influence. In such case it would serve him right to "Court" him. Another reason why it is proper to get your fee in advance is that many would never come and pay it till you had sent them a bill by your collector, and would then indignantly claim that you had insulted and exposed them by sending a bill of that kind.

Also, when at all convenient, get your fees in advance for transient attendance on persons injured in bawdy-house fights, drunken buggy-rides, soldiers, sailors, and the like.

At the same time, bear in mind that you have no right, either legal or moral, to expose the nature of any person's disease to any one, on account of his having failed to pay your fees, even though it was gonorrhœa or he was covered from the crown of his head to the soles of his feet with syphilis.

Venereal diseases are the result, generally, not of providential misfortune, as are other inflictions, but of voluntary indulgence in vice; therefore, self-inflicted. And for this valid reason such venereal patients have not the same natural claim upon your sympathy as other sufferers. In all cases of this kind try to get a just, remunerative fee before you undertake the treatment; then honestly do your duty to the patient until he is cured. Having paid you, he is not likely to change from you to another, and should his case proceed slowly he cannot then suspect that you are purposely running a heavy bill on him, or delaying the cure on account of his being a good-pay patient, as he might do if he were paying you a dollar or two for each consultation.

Many men imagine that they cannot be suffering from constitutional syphilis unless they have detected a terrible chancre at the beginning; and you will often experience a difficulty in making persons who have not detected a primary sore believe their case to be syphilis. Some men will actually stare, scan, and quiz you when you tell them they have the p–x, as if they thought you a quack or impostor trying to frighten them out of money. If you can show such a patient a fac-simile of his chancre, roseola, or mucous patches in your text-books on venereal diseases, or even read with him a description of them, it will awaken him to his real condition and put him on his guard against either neglecting his case or infecting others.

When you feel certain that your diagnosis of syphilis is correct, look the patient in the face, and, with a manner that indicates your practical knowledge of the matter, tell him that in your opinion he has true syphilis, and be careful not to be browbeaten into taking charge of the case for a trifling fee. It is a grave disease, and the responsibility and worry of the medical attendant are often very great and protracted; the fee, therefore, should *never* be nominal.

You can readily broach the fee question to any patient suffering from a private disease by remarking, immediately after making your first examination, "Well, I see what your case is, and am willing to take charge of it and give you my best services, *if my terms will suit you.*" This will necessitate his asking what your terms are, and will afford you the opportunity to tell him. Or, if you regard the services likely to be required as important and valuable, whilst he evidently thinks the reverse, if you will incidentally begin with the remark, "Ah! I fear my charges will be more than you would be willing to pay," this also will compel him to question you upon the subject, and that, too, in a somewhat more favorable frame of mind for your purpose.

Some people labor under the impression that physicians are public functionaries, and that the law compels them to answer the beck and call of any one who chooses to send for them, pay or no pay. *It does not;* you have a perfect right to refuse for any reason that is satisfactory to yourself; but your time is supposed to belong somewhat to your suffering fellow-creatures, and you are expected to be ever waiting and watching in complete readiness; and both the profession and public opinion would severely judge and condemn you if you were to refuse to attend an urgent case to which common humanity should prompt you to go,—especially if you refused on account of fees, and particularly if other physicians were not easily accessible. If you are really "*too busy,*" or "*not well enough,*" or are immersed in another engagement that cannot be set aside, or have another equally urgent duty to perform, these will generally be regarded as sufficient reasons, and protect you against argument or criticism. But "*I'm just at dinner,*" "*I'm too tired,*" or "*I need sleep,*" or "*I am afraid I will be dragged into court as a witness,*" etc., look like a hard indifference, and are not accepted by the public as adequate reasons for refusing to go, and in cases of urgency should never be offered. In the name of Jupiter, what business has any physician to be at dinner, or sleepy, or tired, while yet young enough to crawl, or with strength enough left to think a thought, or hold a pen, when the sick public give a call or whistle?

A few persons also believe there is some law or rule that prevents a physician from attending his own wife and children, or other near kinsmen, when they are sick. This belief has been created by the fact that some esteemed brother-physician is generally intrusted with such cases through a fear, in the physician's anxious mind, that personal interest in those so near and dear to him might warp his judgment, or in the event of fatal issue might leave a deep and lasting regret in his mind that this, that, or the other line of treatment was not pursued instead of that which was.

After your work in many a case is done you may have to

"Assume the cloak of necessity to save the fee,"

and use this, that, or the other stratagem to get your fees. Not only should you send your bill to a patient in due time, but if you fail to hear from him within a reasonable while, emphasize it by sending another, with the same date, etc., as the first, marked "duplicate," or "3d bill," "4th bill," as the case may be; for he may not have received the first, or may have thrown it aside with a Tra-la-la-la! or may be purposely neglecting it in the hope that you will cease your claim forever, or trying to let it stand over till it is forgotten or is out of date.

An *effective* plan to adopt with a certain tardy class of patients, when you are in need of money, is to ascertain the date at which you will have a debt or note to pay, or will have to raise money for any other special purpose, and then to write a week or two before the time and briefly inform them that you will have a *special* need for money at the time specified, and ask them kindly to pay you on or before that date. Most people of any worth will exert themselves to comply with the request, if courteously made. In this manner you can well approach both your best and your worst patients, and some that you cannot successfully approach for money in any other way. A request so conveyed, moreover, shows that you do not want merely to get it out of their pocket into your own, but that you ask for it because you really happen to need it. One who is in debt has always a legitimate excuse for sending in his bills as soon as his patients recover.

Another plan, good to pursue with those who habitually throw bills aside and neglect to pay them, is to send your accounts some day when you are in need of funds, with a brief note asking them to pay in the course of the day, and assign your reasons for making so pressing a request. Even though they pay you nothing then, knowing that they have disappointed you in your dilemma, they will feel impelled at least to pay something on the account when they again need your services.

Also, using the phrase on your bills, "Amount now on the books $———," or "Balance still on the books $———," and inclosing a brief note with the bill of a delinquent for whom you are tired of waiting, telling him that his account is greatly overdue, and asking him kindly to call and settle, as you are anxious to close the account "on the books," remind him of the fact that it is "on the books" and overdue, hence probably seen and thought over by you daily, and may arouse him to the extent of calling to pay, or to make some definite arrangement.

By letting your prompt-paying patients know in some way or other, at the visit preceding the final one, that your next visit will be the last that you deem it necessary to make, it will serve as a gentle hint and afford them an opportunity to prepare, and will greatly increase the chances of your being paid *cash* at the last visit. Convalescents from severe illnesses who are told to pay you a visit at your office when able to walk out again, in order that you may see how they are getting along, are very apt to broach the subject of your fees, and either then pay or make some definite promise before leaving.

You cannot put all classes of bills on the same footing; there is *one* class of patients whose bills had better be sent by mail, *another* to whom they should be taken by your collector or other person, *another* to whom you had better deliver them yourself, and *a few* promptly-paying patients whom you had better allow to ask for them. A careful study of these facts will be of essential assistance to you.

Items and details are, as a rule, better omitted in professional accounts, unless specially asked for, inasmuch as they tend to dissatisfy people, and lead to criticisms and disputes that would not arise did not the items furnish a pretext. Assume the position that he who confides in you sufficiently to put the lives and secrets of himself and family in your keeping should feel sufficient confidence and gratitude to intrust you to say what value you deem mutually fair to place on your professional services. In fact, a physician's bill that gives in detail the various *items* is more apt to be disputed or criticised unless it be unjustly small. Bills that simply state the *total* amount, or "amount due for services since date of last bill," or "amount now on the books," are much more likely to be paid without dispute. The items, however, of every bill should be carefully entered in the ledger, in order that the charges may be verified if requisite; and each and every charge should rest on a distinct financial base of its own. Should a patient question the accuracy of a non-itemized bill, at once concede his right to be furnished with a statement of the number and dates of visits and any special services charged for, or permission to see the items on the ledger should be permitted or suggested. But few who would intrust you with their lives would push you to this extent after serving them faithfully, and these had as well be erased from your list of patients.

On the payment of money other than a simple cash fee by your patients, it is well to insist on giving receipts, even though they should deem it unnecessary. Compelling every one who pays a debt that has been booked to take a receipt not only prevents subsequent disputes, but assists also in maintaining a regular and desirable business-like system between you.

Be especially careful to avoid soul-narrowing avarice in its various forms—meanness, greed, oppression, stony heart—and all other hateful extremes. If you attempt to shave too closely in money matters,—when a patient is so low that it is no longer decent to take fees, or hungrily hold watches, jewelry, or other articles as security for the payment of your fees, or compel their owners to pawn or sell them for your benefit, or charge interest on your bills because not promptly paid, or be unreasonable (Shylock) or too vigorous in your efforts to collect fees from any one,—it would not only be morally wrong, but would be very apt to prejudice your reputation and create a feeling of hostility against you that time could not efface.

For a like reason it is, as a rule, better not to charge for a certificate of sickness furnished to patients to enable them to draw sick-pay from clubs and other beneficial societies, or for school-children's certificates of vaccination, etc. These should be regarded as personal favors, differing from cases in which a fee is right and proper.

But in every case requiring you to go and make an affidavit before a court or magistrate, a moderate charge is proper.

It will seldom pay you to sue people, even though your suits be successful; indeed, it is, generally speaking, undesirable for you or any other physician to begin litigation to enforce your claims, except under very aggravating circumstances or to maintain your reputation or self-respect. Physicians who frequently go to law to recover fees generally lose more in the end than the yield, by exciting prejudice and making enemies. You should never resort to compulsory measures with any one whose failure to pay is due to honest poverty. While naturally seeking to get good patients, who can and will pay for your services, be ever willing to do your share of charity for the deserving poor; at the same time the necessity of earning a living for yourself should make you careful not to let it crowd out your remuneratory practice.

When called upon to attend cases of sudden death, drowning, suicide, persons found dead, murder, etc., in which the unfortunate victim is dead before you can get to him, or in calls of emergency, where another physician reaches the patient and takes charge before your arrival, or in other cases where your services are not called into action, or are merely nominal or clearly useless, it will, as a rule, be wise not to send in an account, as under such circumstances not only would it generally be left unpaid, but be harshly criticised. If, however, a feeling of gratitude induce the people interested to tender you fees, for your trouble, accept whatever is right.

In obstinate and invincible maladies, such as hopeless cases of cancer, phthisis, aneurism, etc., in which, after having gone the rounds of the profession, you are consulted in the very last stages, with the hope of getting a new heart, or a new pair of lungs, or having other miracles performed; or merely to see whether you can possibly do anything of benefit to them, you had better deal candidly, and frankly acknowledge that you can do but little, or nothing, and decline the fee *even if tendered*.

It is better, as a general rule, to make *no charge* for ordinary or trifling advice incidentally given to patients when they call to pay their bill, or to persons for whom you happen to prescribe in public places (curbstone prescriptions), where you are *not* pursuing your professional avocation. Such exactions would, to say the least, tend to engender unpleasant reminiscences and harsh criticism. Every physician occasionally writes prescriptions under circumstances that, even though he be technically entitled to remuneration, *his own* interests forbid his charging or even accepting a fee when tendered.

Never make a charge where the fee would come from another physician's pocket; every physician attends his professional brethren and members of their families gratis. Some also attend clergymen and their families without a charge of money, especially those with whom they have church relations, and those who receive salaries so meagre as to make the payment of medical fees a hardship. But where a clergyman is in the receipt of a liberal salary, and his calls on you are frequent or onerous, I know of nothing in ethics to forbid your accepting from him a fee voluntarily tendered. Some of our best physicians make it a rule to charge half-fees to their own spiritual advisers; that is, they make out the bills for the full amount and receipt them upon payment of half the sum. Their influence, if properly directed, is supposed to cancel the remainder.

Never oppress any one by exorbitant fees. Nearly every one depends on his physician's unwatched integrity, believing that he will be honest in his conduct, honest in his treatment, and honest in his charges. Be especially fair in your charges against estates, and in all other cases where unusual circumstances place the debtor at your mercy. These opportunities will fully test whether true honesty has a seat in your heart.

"As a man thinketh in his heart, so is he."

When you are in doubt what to charge, look upward, then make out your bill at such figures as you may deem just to the patient, to the profession, and to yourself, and thus show clean hands, morally as well as antiseptically. Even-handed justice is the basis of all lasting reputation.

Great injury is inflicted on our entire profession when Dr. Chiselum, Dr. Tinchaser, Dr. Highprice, Prof. Twentyfold, Dazzlefee, or any other of our guild places an exorbitant value on his time and labor, and charges those whom chance has placed in his power a fee so enormous or outrageously extortionate as to cause great gossip or newspaper notice of it. But, carefully avoid making censorious or derogatory comment, in the presence of non-professional persons, on the fees claimed by another physician, unless you are fully acquainted with all the circumstances, for he may actually have good and sufficient reasons for the charges made.

When you and a professional brother do each a portion of the work in cases of accident, confinement, etc., a very fair plan is to agree to charge a joint fee and divide it. When you receive such a joint fee, go at the earliest possible moment and divide every dollar, fairly and squarely, with your fellow-worker, on whatever basis you have agreed upon.

When another physician is called to a case of yours, during your absence, not only thank him at the first opportunity, but also insist on his sending his bill for whatever services he has rendered. No one can be expected to work under such circumstances without fee. His kindness to you consists in having responded to the call.

Never acknowledge or work under the job-lot fee-table of any association or company, *unless* it be in harmony with the regular professional fee-table of your community.

A fee-table should never be extravagantly high on one hand, nor meanly low on the other, but should be reasonable in its tariff, and should always allow a reduction if the patient's circumstances require; and should also allow attendance on the moneyless poor gratis.

Humanity requires you (as God's instrument) to go promptly to all cases of sudden emergency, accidents, and the like, in which the life or limb of a fellow-creature is in jeopardy, without regard to the prospect or otherwise of a fee. You should do various things for the sake of charity; among these is to give relief to any one injured, or in great pain or suffering, regardless of fees. At such times regard only MAN *in distress;* show no distinction between rich and poor, high and low, but consider only your simple duty to suffering humanity. The good Samaritan succored the wounded man, took him to an inn, and provided for his immediate necessities. You, as a physician, should be equally humane and prompt to go and bind up wounds, and relieve suffering in all cases of emergency. After this is done further attendance is, of course, optional, and depends upon whether you choose to render it, or feel that you can afford it; but you are really no more bound to continue to attend such a one gratuitously than the baker is to give away his bread to the hungry, or the tailor to give away his clothes to the ragged.

But, take care never to slight the worthy poor, who are under the iron heel of poverty and need medical attendance. To the poor life and health are everything; their very poverty and lack of comforts make them more likely to get sick and to suffer more in sickness than the rich, and worthy kindness to them in worthy ways should be as broad as God's earth. Besides, there are none so poor but that they may amply repay your services by their earnest "God bless you, Doctor," and their genuine, lasting gratitude. Besides, how heartfelt and pure the gratification to wrest a fellow-being from destruction!

Physicians render more gratuitous and unpaid services than

any other class of people in the world. Allowing that there are in the United States fifty thousand regular practicing physicians, and that each does one hundred dollars' worth of labor to charity practice a year,—which is far below the average,—we have the enormous sum of five millions of dollars of charitable labor given by its medical profession every year.

"The poor," said Boerhaave, "are my best patients. God will be their paymaster." But even in dispensing charity, careful discrimination is essential. There would seem to be three classes of the poor,—the Lord's poor, the devil's poor, and the poor devils. The first and last are worthy objects of every physician's attention, and you would do well to lose no opportunity to give relief to their ailments. The less, however, you have to do with the other class (*the devil's poor*), and the less health and strength you waste on them, the better for you; nevertheless, you will be more or less compelled to attend more than you would otherwise care to do of the lowest and vilest victims of vice, intemperance, and sensual indulgence,—who are perhaps a curse to their families and a nuisance to the neighborhood,—and watch over them as faithfully as if they were noblemen: some for God's sake, and others, it may be, on account of their relationship to better and more provident patients; you will generally find, however, that, "though this citizen and that fellow may be brothers, their pocket-books are not sisters."

It is your duty to raise your voice in the profession against the fearful abuse of medical charities by the people, and the largely increasing numbers of free special dispensaries, college clinics, and the out-door departments of hospitals, church infirmaries, and private retreats, which, of late, under the color of *charity*, attract not only aching beggars from squalid streets and alleys, drunken and worthless men's families, the poverty-stricken sick and humble people out of employment, whose forlorn aspect is unmistakable, but also thousands of stingy impostors and miserly drones, who are *abundantly able to pay* for medical services; and, which, still worse, offer a refuge in their rainy day to the lazy and vicious, against which, the latter, consequently, need not provide by industry, sobriety, and economy. Make a person a pauper, or encourage him to become a lazy beggar, or destroy his independence and manhood in one thing, and he is apt to degenerate and become improvident and worthless in many.

No member of the profession—and the same may be said of pharmacists and physicians who keep drug-stores and prescribe over their counters—has, in the spirit of common justice, a right to give professional services to the public without fee, except to the moneyless poor (to whom they should be rendered in the holy name of charity, as freely as the air they breathe); for, although there may be no loss thereby to him personally, it has a pauperizing tendency on a certain class of people, and is taking bread from the mouths of struggling physicians by monopolizing practice that would otherwise fall into their hands, and to that extent it is despoiling the profession of its legitimate fees.

> Glory built on selfish principles is shame and guilt.

Thousands of young and deserving sons of Æsculapius have been, of late, cheated out of what would be to them bread and a slender support, and a chance to get into practice, by so-called "Hospital" or "Church" Charities, carried on chiefly in the interest of individuals, or coteries, who, to foster reputation in their specialties, and to outstrip rivals, treat *everybody* that applies,—the rich, the poor, and the intermediate class,—whether entitled to the benefits of their charity or not, without the slightest regard to the interest of other medical men, or their desire to do a share of charity.

> Immortal gods! Such stony injustice
> Blots all the heaven-born features.

The ultimate result of this state of things will be either that the profession will, in self-defense, be compelled to organize *self-preservation associations*, or that individual physicians will take up the case and resolve neither to turn over cases to nor to call into consultation any specialist, professor, or surgeon who continues to render gratuitous service to those who are able to pay for it. The last-mentioned course would probably influence the transgressors strongly. THE SHAMEFUL WRONG DONE to the profession by such institutions lies not so much in the working of the hospitals themselves, but IN the conduct of THEIR DISPENSARIES AND OUT-DOOR DEPARTMENTS.

Probably a considerable proportion of the impostors and frauds able to pay for services, who impose on these institutions, knowing the risk of being unearthed and turned away, would shrink from venturing such exposure to the public by the prominent display of some such sign as the following: "This Dispensary is for the moneyless poor only."

Bear in mind, for an individual to advertise gratuitous attendance on the poor at his office, or at certain times, or under certain conditions, is unprofessional.

Found your ideas of Christian duty and of doing charity on the fifth, sixth, and seventh chapters of Matthew and the thirteenth chapter of First Corinthians, and you cannot go far astray.

"Prompt payments are appreciated by everybody" is a very useful maxim to have printed on the margin of your bills; it is truthful, and gives thanks to those who pay promptly. To those who do not it serves as a neat admonition.

The size of the house does not always show the size of the owner's honesty. You will find, in the course of your professional career, that honesty and dishonesty are not confined to any one nationality or to any station in life, but that there are many very good men and others equally bad among the rich and poor alike. You will, perhaps, mount many a marble step, pull many a silver bell-knob, and walk over many a velvet carpet for well-housed, sumptuously fed, fashionably clothed, diamond-studded patients,—

"With the manners of a marquis,"—

who will turn out unscrupulously fraudulent, and at the same time you will get many an honest fee from others who make no great pretensions and possess but little save their truly honest hearts; it will touch you, to see these come with part of their small pittance to share it with you. Others, who know what it cost to get what they have, know how to hold it, and the demands of fashion are now so great on those who are trying to keep up with it, that many with moderate incomes habitually ignore their physicians' bills in order to aid in keeping up appearances of being worth more than they are. You will see many a man bowed down with debt and despondency, while his trinketed wife and dazzling daughters parade about as gay and as fine as strutting peacocks, indebted to everybody and paying nobody. Artful, double-dealing women will sometimes actually intercept your bills and make it impossible for you to solicit payment from their husbands, unless you resort to strategy and get your bills delivered direct to the latter; and will even then enter the field of falsehood and do everything they can to defer or altogether prevent payment.

Families will occasionally conceal from the person who holds himself responsible for your bill the true amount of service you have rendered, or the actual number of visits you have paid, and thereby lead him to think you have charged very high, or even exorbitantly. Be prepared, therefore, promptly to correct such errors.

The most unsatisfactory and troublesome kind of patients physicians have to contend with are the *unprincipled tricksters*, who, wholly void of moral sense, cheat everybody that affords them a chance, and consider it only an honorable transaction to victimize physicians, and would not cross their fingers to keep us from going to the almshouse. You will be fortunate if you have sufficient tact to avoid having anything to do with those who belong to this class. It is far better courteously, but firmly, to decline to accept as patients those who can but will not pay, without assigning any reason, except that you are "*too busy,*" or "*I'd rather you would consult someone else,*" than to have to wrangle with them about your fees after your work is done, and maybe, after all, get neither fees nor thanks.

Have your wits about you, and tell Hardnut, Spendall, Dedbroke, Poormouth, Bluffum, Codfish, and other habitual delinquents, who have plenty of money to smoke expensive cigars, go to places of amusement, buy beer, or fill the brandy-bottle, or to furnish their houses like palaces, or to follow the follies of fashion, but *none* to pay the physician,—when they have the temerity to come, with lamentations and a hatful of excuses, to increase their indebtedness,—that they are already as largely indebted to you as you can afford to let them be, but that you are perfectly willing to serve them again *after* they pay you what is already on the books, or a reasonable part of it, or if they will pay you for the new services cash at each visit; and base your position in the matter not so much on the fact that they are in question, as that you are acting in accordance with a regular rule. Such attitude on your part will very probably lead to some more or less satisfactory action on theirs, and thus indicate to you what course to pursue.

In dunning delinquents for fees, it is better to charge them with carelessness in the matter of paying, than with dishonesty.

You will encounter many a person who, although quite amiable during your attendance, will prove very different— maybe as sensitive as the eyeball—when your bill is presented; then,—

> "Oh, such vinegar aspect!"

In such cases, take especial care to give no cause for fault-finding with your mode of presenting it. It is a useful precaution to inclose each bill sent by mail or messenger in a half-sheet of blank paper, so as to prevent prying custodians from peering through the envelope and recognizing its contents.

When possible, let your bills be presented direct to the party financially responsible, or to the real head of the family, and say nothing about them to other members of the household.

In spite of all, were you a Solomon and an Angel combined, many patients will find fault, show ill-temper, and meanly quit you, under one pretence or another, when you send your bill or ask for your fee, no matter how or when you do it.

A moderately successful practitioner has about two thousand persons who call him *their* " doctor " (fully three hundred of whom are moneyless or bad pay); and whenever any one of these is suffering from mental or physical ailment, he must share it by head-work and hand-work and heart-work. He must combine all good qualities, and appear the perfection of each to all men, must be bold as a lion with one patient, as patient as an ox with another, and as gentle as a lamb with the next. Self-sacrificing, his own aches and pains must be concealed or go unnoticed,—

"It is a fortunate head that never aches,"—

and, being the slave of the sick public, he must face contagious disease and inhale noxious vapors, miasms, and malaria; encounter the filthiest kind of filth and the worst of all stinks, and perform many distasteful and disagreeable and disgusting duties, amid embarrassments, disappointments, and vexations.

"None but a physician knows a physician's cares."

He must endure all temperatures,—August suns and December blasts; drowned with the rain and choked by the dust, he must trudge, hungry and sleepy, at noon or midnight, while others, oblivious to care, are resting, or being refreshed with sleep; must be with families at all seasons, in death and recovery, in sorrow and joy. A soldier may serve his whole term without smelling powder or even getting within long range of danger. A physician is in continual danger, and when, like a wild and relentless tornado, the swift, gaunt, ghastly, withering epidemic begins its work of death, no matter how great the danger, he cannot flee but in dishonor,—no personal considerations, no domestic relations, no plea whatever can excuse him,—but he must depend on Providence, and, from pure love of humanity, take his life in his own hands, hazard the danger, and stand (like Aaron) between the living and the dead, in localities filthy and ill-ventilated, to fight the monster —

"With aspect stern and gloomy stride"—

face to face, even though, without reward or expectation of reward, he suffer martyrdom in the conflict, while thousands are falling, like sheep, around him, and other terror-stricken thousands are fleeing for their lives! He must have an eye like an eagle's, a heart like a lion's, and a hand like a lady's, —must combine all good qualities, and appear the perfection of each to all men, and, heaven knows! from the narrowness, and crookedness, and steepness, and roughness of his life's road, he deserves far more generous treatment, and a much more comfortable support, than he receives.

Some one has divided man's life into four periods, and called the first twenty years the period of preparation; from twenty to forty, the period of struggle; from forty to sixty, the period of victory; and after sixty, rest. No fourth period for the physician; his struggle lasts (if he is able to walk, to see, or to hold a pen) until his life ends.

How nice it would be if a physician could retire, with honors and a competence, at sixty, and leave the path open for other and younger men!

"A youth of labor with an age of ease."

Computed by the ten-hour system, every busy physician does no less than five hundred days' work a year, loses much sleep and many meals, and has to serve numerous masters at all hours, from sunrise to sunrise. Every year, measuring by work, vexations, anxieties, discouragements, and care, the average practitioner has three years of brain-work and mental strain, has to endure all kinds of criticism, does more charity, and then lets his accounts against those who are able to pay run longer than any other person in the whole community.

The trades and common occupations are learned in three or four years; perfection in them is then reached, and the balance of life is simply a routine employment; not so with us, for in medicine the law is progress, perfection is never reached, and study and mental exertion are never done.

New discoveries teach new duties.

The fact that a physician has to keep up an external show of prosperity, and that many pay their visits with gloved hands and in stylish carriages, leads not a few unreasoning persons to infer that ours is a path of ease, almost a bed of roses; that we drive about during bank-hours, prescribe for a few select patients, receive fees by wholesale, and soon get rich enough to retire and live on the interest; all which is a very great mistake. On the contrary, every older physician knows that after working hard day and night, owing to the difficult collections and the large proportion of the poor, the practice of medicine is neither an Eldorado nor a money-making profession, and that it is almost impossible to get rich by the practice of medicine, unless one have extraordinary professional skill and repute, or be a celebrated surgeon, commanding great fees; or a fashionable favorite, lucky enough to attend groups of patients who have copious and open purses, or a leading specialist, charging what he pleases—

"Their hens lay eggs with double yelks."

In fact, I know of no legitimate business in which the same amount of capital and time laid out, and labor, industry, and prudence exercised, would not be likely to prove much more lucrative. Other men,—the farmer, the merchant, the mechanic, and the artisan,—successful in their pursuits, can increase their business to any extent by employing additional hands and superintendents. A physician does nothing by proxy, and must undertake no more than he can do personally, and has no gains but from his own individual efforts. Besides, the expense of living and the cost of library and apparatus have all greatly increased within the last few years, and the fees for services have certainly not advanced in the same ratio.

The income of the most successful physicians is far below what is commonly imagined, and many a physician is in a constant state of poverty and debt, even after economising in every direction and foregoing the purchase of many books and instruments which he actually needs. Besides, ours is not a long-lived profession, and many a conscientious, able, time-worn physician dies, and, instead of bequeathing an Aladdin's lamp, leaves those dependent upon him poor and helpless, unless he has acquired money otherwise than by his practice.

After his death, a physician's outstanding bills are rarely collectable. Many a one with a large practice dies, his poor family inherits only a book of worthless accounts, and his estate is found to be scarcely worth administering on; as if they had spent their lives in

> "Dropping buckets into empty wells,
> And growing old in drawing nothing out."

According to the mortality tables, the average of the lives of physicians is fifty-six years. If you begin practice at twenty-four, your active-life prospect will be thirty-two years, and from a thousand to fifteen hundred dollars will represent your average yearly income.

> "Facts are stubborn things."

Now, were you (through God's mercy) to practice these thirty-two years without losing a single day, and collect (say) eight dollars every day of the time, you would receive but ninety-three thousand four hundred and forty dollars. Deduct from that amount your expenses for yourself and family, your horses, carriages, books, periodicals, and instruments; your taxes, insurance, and a multitude of other items for the whole thirty-two years, and then, so far from being rich, even after this long and active life of usefulness in our important and honorable profession; yea! after a whole life-time of scientific work, mental toil, and of slavery to our unrelenting taskmaster, The Sick Public; from the days of the dirty, unwholesome dissecting-rooms through all life's phases to old age; with not even the Sabbaths to call your own,—when your harvest is past and your summer is ended you will have but little, very little, left to support you when you reach the down-hill of life, or are broken down in health, with memory worn out, eyes dim, arms' strength and hands' cunning lost, other faculties deteriorated, unfit, unable to work, and in need of a physician yourself.

> "Thus they who reach
> Gray hairs die piecemeal."

The physician is, as a rule, so poor a man of business that if he receives money enough to meet his necessities he is but seldom troubled about the balance. Money comes, money goes, and he saves nothing. The writer had a friend, a strong man and an excellent physician, who detested keeping accounts, and was so neglectful about his fees that he kept no systematic register of charges and payments whatever, trusted all to his memory, and rarely sent a bill; the result was that his easy and convenient terms, together with his superior skill, made him extremely popular, and brought him more business than he could do justice to, and kept him overworked day and night, until, at the end of fourteen years, the incessant fatigue, exposure, anxiety, and crowding cares of his overgrown practice ran him off his legs, broke down his giant strength, and he died, almost, as it were, by suicide, leaving his starving wife and unfed children without a dollar—yes! nothing—except painful regret at his improvidence and lamentable lack of business system. He was, indeed, the "pet of the town" while he lived; but how fared his wife and children after his life's work was over?

Be it your duty to self and to others to guard against such a system, or, rather, lack of system; for, while you owe certain duties to your patients, you also owe some to yourself and some to your family, if you have one, and no man should ever sacrifice and neglect either department for the other.

One would suppose that physicians, whose lives are spent in preventing and curing disease in others, might themselves claim exemption from disease and decay; might turn aside from their own bosoms the arrows which their skill has turned aside from so many others, and attain unusual longevity; but not so. On the scroll of the Icy King of Terrors we are but men like other men, and have no exemption from the common lot; are bound by the same laws of mortality, and, subject to perpetual wear and tear of body and mind; we suffer sickness, we are deprived of health, our bosoms receive the shaft, and we pay the natural debt, and fill an early grave fully as often as other men.

> "Death!—great proprietor of all—
> Will seize the Doctor too."

Remember that other business-men's resources and productiveness survive their death or outlast their ability to work, while a physician's gains represent nothing more stable than his individual capacity for labor, and end when he does; therefore, while you are young and healthy determine to put away part of your income as a nest-egg for a rainy day, or to fall back on in sickness, or when old and tired of occupation; for no one knows what ill-luck may overtake him in the course of life, or how dire may sometime be his need for money; furthermore, even if one is lucky enough to remain healthy, it is the dollars saved during the first years of practice that roll up into future competence.

> "For age and want, save while you may."

Besides, if your death would leave your loved ones otherwise unprovided for, it would be wise and reasonable to take time by the forelock and provide for them by a sufficient assurance on your life, which can be gotten and maintained at a small cost; then, if you should be taken,

> "The widow's heart shall sing for joy,
> The orphans shall be fed."

Beware of investing your earnings in popular speculations, and refuse to go security for other people's debts, etc. Physicians are notoriously unfortunate in such ventures, and they have caused many of our number to end their days disappointed and moneyless, instead of in comfort with a competence.

A good, honest collector—one who possesses judgment and sufficient tact to wake up hard customers and get money on an easy installment, or other plan, from reluctant and dilatory debtors without irritating and converting them into active enemies —will be found very useful, and is quite necessary if you be too tender or too high-spirited to allow a direct transfer of remuneration from old friends or refined patients, or if you have no time, or are an indifferent collector yourself. Having only business transactions with patients, his interviews with them are *business exclusively*, and he can persevere in his efforts to collect to a degree that you would find unpleasant or humiliating. Many thoroughly honest people are too poor to pay large bills, and if you allowed their account to accumulate from time to time into a large bill they would be unable to pay it, even if they wished, and consequently you would place them in a position of embarrassment. Having a collector prevents this and keeps one's financial department in a healthy condition. It also tends to stimulate those who are habitually slow of payment, and, at the same time, sifts out undesirable patients and erases their names from your list before they run their bills very high.

You should have some specific agreement with your collector, not only in regard to his rate of percentage for collecting, but also as to the conditions under which he is to claim it. Among other things, you should stipulate that he is to make full returns to you once a week, or, at least, once a fortnight; that he is to have no percentage on money paid to you by those whom he has not visited for a month, unless you have at their request stopped him from calling; and that he is to receive nothing on bills placed in his hands if the indebted parties call and pay before he has delivered their bills; in fact, nothing on any bill which he does not in some way assist in collecting.

It is wise to post your books, make out bills, settle with your collector, and, in fact, to conduct all the features of your pecuniary department as much out of public sight as possible, so that the public may know little or nothing about you except as a medical attendant.

If you adopt some special shade or color for your bills, it will not only make them easy to find when patients mingle them with others, but will also remind those who are remiss or tardy in paying the debt, every time the color arrests their attention, and may, by thus constantly reminding them, actually secure or accelerate payment.

The publication of lists (black-lists) of the names of fraudulent patients among physicians practicing in a given area is mutually profitable, as it is a means of debarring those who can pay if they wish from systematically imposing on a succession of physicians, and coercing them into paying and retaining some one. From such lists the deserving poor, unable to pay, should always be omitted.

A good way to get up "The Physicians' Protective Alliance" is to have a meeting of the physicians of your section, and, after organizing, appoint a Publication Committee, to which every member shall, within a specified time, hand a list of the names, occupations, and addresses of able-to-pay patients who have, through apparent carelessness or lack of good principle, owed them bills *unjustly long*.

All these names should be alphabetically arranged and published, in a small, plain, blue, cloth-bound "Reference Book," one copy for each member. Also, have to accompany each book a *separate* printed slip, containing the name of each physician who has given a list, with the number assigned to him by the committee placed before his name:—

    1. Dr. John Allen,
    2. Dr. Henry Blair,
    3. Dr. William Curry, etc. ;

these slips to be kept sacredly private, and seen by their owners only. Suppose Dr. James Shaw is No. 16 and Dr. Thomas Wilson is No. 31 on the slip or key. We find among the delinquents the name of Samuel Adams, plasterer, No. 127 N. Bond Street, with 16 behind it. This, of course, shows that Samuel Adams has been careless or unjustly slow in paying No. 16 (Dr. Shaw) a bill that he owes. If 16 and 31 both appear behind his name, it shows that he is in bad standing with both Drs. Shaw and Wilson, and has been reported by both. The object of such an association should be: *not to forbid* any one who chooses to attend to delinquents from doing so, but simply to tell one another of them, so that any one may either decline to attend them or do so with his eyes open.

The list of names in the book should, for obvious reasons, follow some such inoffensive title as:—

THE PHYSICIAN'S PROTECTIVE ALLIANCE.

"BUFFALO, N. Y., January 1, 1893.

"THE FOLLOWING IS A LIST of persons who, through apparent carelessness or lack of just principle, have been indebted to various physicians *unjustly long :*—

" Adams, Samuel, plasterer, 127 N. Bond Street, 16.

" Bowman Daniel, engineer, 479 W. Biddle Street, 23, 44."

Every two or three years a new volume should be gotten up and issued.

# CHAPTER XII.

*"The more one believes in the possibility of error, the surer will he be to avoid mistakes."*

BE alert, observant, and apprehensive. You will be supposed to foreknow all conceivable things relating to disease, its dangers and its terminations; therefore never exhibit self-accusing surprise at any possible event growing out of sickness. Even when cunning death has unexpectedly visited some one under your treatment, either directly or as a coincidence, do not let your manner or expressions indicate that you were altogether ignorant of its possibility, or that you regard yourself as deserving of blame, since every case has not only its probabilities, but also its possibilities.

When you are attending cases in which there is danger of rapid or sudden death, beware of ordering chloral, opiates, or other potent drugs in such a manner as to create a belief that they have caused or hastened death (manslaughter). Circumstances or fear of coincidence may at times even render it judicious to avoid writing a prescription at all, and simply to order this or that appropriate remedy under its common name, so that, its suitableness to the case and its innocuous nature being understood by all, you may not be unjustly charged with doing harm with it.

When any one under your treatment sinks unexpectedly, or dies mysteriously, or shortly after the use of some agent that you have directed, or after the administration of some new remedy, or shortly after you have performed some operation, or soon after you have pronounced him better,—

*"Joy and sorrow are next-door neighbors,"*—

or in any other way that could possibly subject you to unjust implication or blame, it is better quietly but resolutely to make a visit to the house of mourning, with a view to ascertain the cause of death, and also to discover what attitude the friends assume toward you, and to meet their criticisms and protect yourself by explanations, etc. On such occasions you cannot be too calm and self-possessed, nor too well prepared to explain, and, if necessary, defend your course and the treatment. By so acting you can anticipate injurious and prejudicial reports and suppress or shape them before they become widely circulated.

On eagle's wings, scandals fly.

Bear in mind that such deaths are often due to gross imprudence of patient or friends, or to some mischievous article of food or drink that has been smuggled in.

Dropping in for the purpose of preparing and giving to the family the certificate of death affords a good chance for a desired interview after any one's decease.

When you are called to a case of sudden death the greatest composure of mind and manner is essential and important; be guarded and discreetly reserved,—

"The tongue is the rudder of our ships,"—

and never assume an oracular or prophetic air, or express any opinion of the cause in any such case, but show a Sphynx-like determination neither to form nor deliver one, until you have carefully collected and duly considered all the circumstances.

"Second thoughts are best."

The possibility of death being due to embolism, or paralysis of the heart, syncope, pulmonary apoplexy, or other disease of the heart, or lungs, or brain ; to poison, violence, or suicide ; should be calmly and thoughtfully weighed before you express any opinion ; for, should you rush in with a flurry, neglect this precaution, and un-call-back-ably christen the disease according to your first-born opinion, further developments in the case may prove it to be some other well-known affection, and expose you either as a butt to pleasantry and ridicule, or to severe censure and deep mortification.

If you are called to a case of sudden death in which violence is suspected, or to which you are summoned by the police or coroner, be very careful to note everything in connection with the body and its surroundings, and also where a post-mortem is necessary, the condition of the viscera, each one of which should be carefully examined before giving an opinion as to the cause of death. Your notes should be taken by yourself or an assistant at the time, in non-technical language, recording first the year, day of the month, and the hour, then the facts of the case and your interpretation thereof, and subsequently your comments. These notes should be preserved; as you will be allowed to peruse them in court, if summoned there to give evidence, in order to *refresh* your memory; though not wholly to rely on them. If the cause is suspected to be poison, be very careful to tie the stomach at both ends before its removal, and keep it and its contents in clean, sealed vessels, under your own eye and custody, till a chemical analysis can be made, unless their care be confided to the police. If a person be dying from the effects of violence (wounds or poison), when called to him, calmly and feelingly impart the fact to him, and if he volunteer a statement of the circumstances causing his injuries, or in reference to his assailants, take his words down at once in his exact language, as such a statement will be received in court as if made under oath, provided the person makes it under the belief that he is about to die of his injuries.

The mottled, reddish, or livid patches, and the purplish-black discolorations which appear on bodies shortly after death occasion no little talk and exaggeration among the laity, and are often cited as evidence of the malignant or putrefactive nature of the death sickness, or as proof of ante-mortem violence, while they are really due to post-mortem contraction of the walls of the arteries, which squeeze the greater part of their blood into the veins; through whose flaccid coats a portion of its separated coloring matter escapes into the surrounding tissues, creating the appearance mentioned. The escaped fluid tends gradually to collect, by the law of gravity in the most dependent parts of the body, as the back of the neck, trunk, and limbs, thus leaving the higher parts clear and wax-like in appearance.

You can always distinguish these post-mortem appearances from bruises inflicted during life by making an incision into them. If post-mortem, you will find the blood-stain superficial and not involving the tissues beneath, but the contrary if due to violence during life. In the latter case, moreover, they cannot be removed by pressure or change in the position of the body.

The popular belief is that if a sudden death begins at the heart there must have been a pre-existing disease of the heart, and the family physician is often reproached for not having discovered it during the patient's life-time. You will do well to explain that the healthiest heart may suddenly become paralyzed or mechanically occluded (thrombosis or embolism) and sudden death result. Bear in mind, also, that the ordinary termination of organic heart disease is not sudden, but very slow, death, preceded by dropsy, inability to lie down, etc.; in fact, with the exception of cases of aortic stenosis, or regurgitation, or fatty degeneration, there are few, if any, forms of organic heart disease that cause sudden death. Of course, syncope, from mental emotion or physical exhaustion, if not promptly and properly met, may cause sudden death, even when the heart is entirely free from disease.

A belief that stout, healthy people endure accidents, operations, accouchements, diseases, etc., better than weaker, complaining people is another popular error. The truth is the latter are schooled to pain, to disordered functions, lack of exercise, etc., and when they have to endure afflictions, the mutation from their ordinary condition is less than in the former, and they have not so much vital force to be perverted into morbid action, and in many instances their cases turn out more satisfactorily. Plethoric systems generally, strange as it may seem, bear depletion by blood-letting, purgation, etc., badly, because their circulation is accustomed to a certain degree of fullness and tension, anything short of which causes disturbance of the different functions. The loss of a few ounces of blood will sometimes cause a plethoric man to faint, while a spare one might have lost a like or larger quantity without injurious effect.

Old persons seldom bear surgical operations well, especially if they have any disease of the urinary organs. Make it a rule, therefore, *always* to examine their urine before operating. If any such patients die from shock, narcosis, hæmorrhage, or sepsis, after your steel-edged interference with harmless growths; or deformities, or ailments which they have endured for years with only a certain amount of inconvenience, you will, in all probability, be greatly blamed, and accused of having operated simply for the expected fee, or to show applauding by-standers your energy, your dazzling skill, or your manual dexterity.

You are not expected to set aside the laws of nature, and will seldom be censured for a fatal issue in the diseases of the aged, and never in those of hard drinkers, or in cases in which you have given an unfavorable prognosis from the first. On the other hand, if a woman dies in her confinement you will be cuss'd and discussed, and if there is any possible chance to blame you it will be done, for the reason that parturition is rightly regarded as totally dissimilar to disease. Child-bearing is designed by nature to increase and not to diminish the number of our race; death, therefore, in labor, which is a physiological function, or during the lying-in, which is a physiological state, seems contrary to nature, and produces a shock, and often evokes severe criticism.

Wretched, heart-broken patients who are suffering acutely, perhaps afflicted with painful, incurable diseases, and the miserable, flabby melancholiacs, with all their emotional chords out of tune, who are a hopeless burden to themselves and to others, will occasionally imploringly ask,

> "Is there no short, no gentler way
> To mingle with our fellow-clay?"

and prayerfully plead to you from the depths of earnestness to give them something to put them out of the (to them) weary, weary, weary world. Likewise, in the case of those who are enduring terrible sufferings from which recovery is impossible, or at the birth of deformed and monster infants, or with helpless imbeciles, the friends will also sometimes hint at, or even openly request, that a sleeping potion may be given with the view to release the unfortunates—by death.

"It were an alms to hang him."

In many such cases you will agree with the view that—were God to take the poor sufferer it would be a blessing; yet with this aspect of the case you have nothing to do. In refusing such solicitations, in sympathetic but explicit language, let your argument be that human life is sacred, and that no man has a right to say another's life is useless, or with Nero,—

"Twenty more with no excuse for living! Kill them, too,"—

and, also, that since a person has no right to end his own existence, he cannot delegate such a right to another, and, even if he could, you would be the wrong person to ask, since your province, as a physician, in the great drama of life is to prolong life, not to shorten it.

So sacred is human life that were you to perform craniotomy and the child be still alive when born, or should you deliver a monster unfit for earth, you have no right to extinguish life in either. You may, occasionally, actually be blamed for saving a life that selfish guardians don't want saved,—whom they want out of the way.

Many cases admit but gradually of a diagnosis and prognosis. In accidents obscure as to nature or degree, and in cases of sudden illness, when you are pressed to say whether you consider the case dangerous, or likely to be of long duration, reply deliberately and avoid giving definite answers, until you see whether any graver affection is hidden behind the present symptoms, whether new symptoms will develop, whether the system will react, and whether there will be a response to the remedies used. During the progress of such cases be careful to school your features and your manner, so that people may be unable to read your hesitations, doubts, and surprises,—

> Like the pages of a printed book,—

and either insist on consultations or, maybe, dispense with your services. Therefore, in giving a diagnosis or prognosis, you should always use the plainest (English) language and as concisely as possible, and, whenever and wherever it is necessary to repeat it, it is best to adhere as closely as possible to the same phraseology.

In cases of accident and injury to people found in an insensible condition on the highways, or lying in bar-rooms or at station-houses, life itself may depend wholly on a proper diagnosis; therefore, although you may strongly suspect them to be due to drunkenness, you will act wisely to do no guessing, but give a *provisional* opinion only, until they return to a sober state. It is better to say, "He is unconscious; whether his insensibility be due to alcohol, or to other causes affecting the brain, it is at this time impossible for any one to say."

Never pronounce that an injured limb is "only bruised or sprained," and order liniment, with assurances that it will be all right in a few days, until you are positive that it is not fractured or dislocated; or the continued pain and swelling may carry the patient to some more cautious physician, who will discover the truth, to his great honor and your great shame. A great many of your brethren have been caught in this trap.

Bear in mind that death following an injury does not always mean that it resulted from the injury.

> "Death has a thousand doors to let life out."

It is well when called to cases of serious burns, cuts, lacerations, fractures, bites, etc., to *mention incidentally* to the family the possibility of the supervention of erysipelas, septicæmia, lock-jaw, etc., and of deformity, or permanent impairment, or whatever other unpleasant results may be reasonably feared, so that the parties may know that you are alive to all the possibilities and probabilities of the case. With regard to burns, remember that the gravity of a burn is often due less to its depth than to the extent of surface involved.

In the course of your professional career you will come into contact with humanity in all its varied aspects and phases, and your patients will greatly differ in the nature and extent of complaint which they will make in detailing their subjective symptoms to you. Some who are naturally stoical and apathetic will fall into the error of *understating* their true condition, fearing that a fuller statement may alarm their friends, or lead you to think their case serious, and to prescribe much and strong medicine for them, or induce you to pay them many visits. Such patients will sometimes die almost without giving a sign. Others, again, of a hysterical or nervous temperament, fearing that you may not consider them as ill as they really are, or as they conceive themselves to be, will, in detailing their symptoms, *magnify* every detail, and seek in every way to impress you and others with an exaggerated idea of the intensity of their sufferings and the gravity of their condition. One of the many advantages which one's regular attendant has over other physicians is his familiarity with these peculiarities of temperament, with the extent of the vocabulary that each of his patients employs, and with the amount of precision which each uses in answering questions and in describing his sufferings. A gilt-edged society lady, a hod-carrier, a lawyer, a backwoodsman, a school-miss, a straight-laced old maid, a sailor, and a girlish dude would each use a different kind of language to express the same symptoms.

In spite of your earnest and best endeavors, you will often be criticised or upbraided for your lack of foresight in relation to the recovery or death of patients. The ability to estimate the vital resistance in each case, by the temperature, pulse, look, visage, voice, attitude, movements, and general appearance of the patient, is essential to the perfection of your skill as a physician. It is something apart from your diagnosis, pathological and therapeutical, and few attain it.

The truth is that life is a *different* quantity in different people: one man will scratch his finger with a pin and die, another will get both legs cut off and live, and you will usually have no other way to judge this or that patient's prospect of recovery from either of the twenty-four hundred different maladies that afflict mankind than by the *average* human standard. You will sometimes have cases which will baffle every method of calculation and surprise you by their possessing a great deal *less*, and others by having a great deal *more*, than the average tenacity of life; and, no matter how careful you are, there exist rocks that are not to be climbed, and pits not to be fathomed, and things which are, from their very nature, unknowable; hence, you cannot, with our present knowledge, accurately and unfailingly prognosticate the endurance power of every patient.

To illustrate what is meant:—

HEALTH, . . . . . . 0.

CLASSES, . . . . . . { 1st. 2d. 3d. 4th. 5th. 6th. 7th. }

Suppose the above seven figures to represent the various degrees of mankind's ability to endure sickness and injury, and that the fourth figure represents the average extent of human endurance power: some patients, then, will actually succumb and die like sheep if the first degree be passed, some if the second be reached, others can endure to the third, and so on, while still others, with iron constitutions, have tenacity of life enough to recover after going as low as the fifth, or even the sixth degree. Now, if you could penetrate each patient's vital recesses and measure, as with the rule and the compass, his assimilation and innervation, absorption and secretion, reproduction and decay, sensation, motion, and reflex action, and the total of *his* endurance power,—could see at what point *his* possibility of recovery ends and *his* dissolution begins,—you could disentangle and unroll the concatenated web of life from perfect health to death, solve the great problem, and make the strength of this web a matter of mathematical certainty. There would then be fewer unanswerable hows and whys, and you would seldom, if ever, be reproached for unpredicted terminations. This neither you nor any other mortal can do; but you can prepare yourself on all points, and make anatomy and physiology your grammar and dictionary, and pathology your crowning study; also, keep your eyes and ears, mind, heart, genius, and talent, all wide open, and make use of the teachings of accumulated experience, and avail yourself fully of every aid offered to you by advancing medical science.

> Full many a pupil has become
> More famous than his master.

Disease and pain and death are parts of the plan of creation. Disease is ever afflicting thousands of earth's children in every clime, while death (on his pale horse) is busy from pole to pole. Fear of the former and dread of the latter are parts of human nature, and these (fear and dread) cause mankind everywhere to employ physicians: the prince in his palace, the peasant in his cottage, and the outcast in his hovel; the citizen in his mansion, the laborer in his shanty, and the felon in his dungeon; the millionaire and the beggar; the conqueror and the captive; the lord and the serf; the sailor and the soldier; the purple of authority, the ermine of rank, and the rags of squalor; the man of religion, the man of law, and the man of science; every nation and tongue, the Christian, the Jew, and the Pagan; the pale-faced Caucasian, the Hindoo, the painted Feejee, the oily and savage Hottentot on the burning plains of Africa; the tattooed, fierce, brutal New Zealander, and the sinewy savage of our own far west; the Esquimau in the blood-chilling Arctic regions, and humanity in the pestilential swamps and jungles of the tropics; wherever sick and suffering mankind is, they turn to our guild for relief.

This reliance of humanity on you as a physician skilled to heal its wounds and to cure its diseases naturally brings you in contact, on one side, with mankind's greatest, most vital interests, and, on the other, with the great science and glorious art of medicine, and makes your power in your legitimate sphere almost monarchial. You go when you please and come when you will, order what you choose and forbid what you may. You are intrusted with secrets that would be confided to no other person, and are as an honorary member and guardian to every family you attend; and you wield strong influence over husbands, wives, children, and servants, and lay down laws to govern each in matters of life and death, and are obeyed almost as implicitly as though you were Julius Cæsar or the Czar of Russia,—

"The foremost man in all this world,"—

and your knowledge, skill, and attention will be many and many a one's last earthly hope.

Thus, you see, no other men under heaven can do as much good as physicians! Others may have the will, but they have not the power and opportunity; this, with its humane nature, makes ours as noble a calling as exists on the face of the earth, —a calling capable of developing all the good qualities of one's heart, hand, and brain.

Bear, therefore, the greatness of your trust and the responsibility and glory and almost divine mission of our sublime and ennobling profession ever in mind, and remember at all times that every action, every phase of your conduct, every word you utter, every look, every nod of your head, tremble of your tongue, quiver of your lips, wink of your eye, and shrug of your shoulders, will be observed and weighed. Therefore, strive to make your character and your methods as faultless as possible, and let no word ever escape you unsuitable to the occasion. Also keep your lamps trimmed and your oil ready, and observe punctuality and system in attending all who place themselves under your care, and strive to do the greatest absolute good for each and every one who trusts to your skill for relief, that you may fill every bosom with kindness toward you, and every mouth with praise; and be truly called A GOOD PHYSICIAN.

---

Thus, my professional brothers, I would attempt to show that the more closely we study the moral and physical peculiarities of the various classes that make up the community, the more clearly we will see that the practice of medicine has a peculiar and complex environment, and that WE SHOULD MAKE SKILL IN PREVENTING, RELIEVING, AND CURING DISEASE OUR CENTRAL THOUGHT AND OUR CHIEF RELIANCE, AND, AS MEN AND BROTHERS, SHOULD DISCHARGE EACH AND EVERY DUTY TO OUR GREAT MASTER'S ENTIRE FAMILY, AT ALL TIMES AND IN ALL PLACES, WITH FIDELITY AND HONOR; and, further, that we must also possess professional tact and business sagacity if we would succeed in the profession to the fullest extent that lies in us, and create for ourselves corresponding spheres of usefulness in the world.

---

IN CONCLUSION, I FONDLY HOPE THAT THIS LITTLE BOOK ON THE PHYSICIAN HIMSELF MAY TEACH THOSE WHO FOLLOW ITS SUGGESTIONS TO SURMOUNT THE MANY OBSTACLES AND DECIDE THE MANY DILEMMAS THAT ARISE IN THE COURSE OF PROFESSIONAL LIFE; AND ALSO AID THEM TO DISCERN THE STRAIGHT AND NOBLE PATH MORE CLEARLY; AND TO FOLLOW IT MORE BRAVELY, MORE FAITHFULLY, AND MORE SUCCESSFULLY; FOR THE BOOK THAT DOES THESE WILL BE OF UNSPEAKABLE BENEFIT, AND WILL LIVE TO SERVE THE PROFESSION FOR MANY, MANY YEARS; AND NEITHER CHISEL NOR HAND — OF BRONZE, MARBLE, OR GOLD — COULD BUILD ITS AUTHOR A BETTER OR MORE ENDURING MONUMENT.

D. W. C.

# INDEX.

Ability, popular tests of professional, 88
Abortifacients, prescribing pretended, 78
Abortion, solicitations to produce, 77, 78, 79
Accidents, demeanor in attending, 174, 324, 330
Accounts, necessity for keeping, 18
Adding suffixes, 91
Advertising, why unethical, 104
Advice, giving, relating to health, 129, 198, 199
Æsculapius and Hygeia, symbols of, 140
Aged, operations on the, 328
Ailments not to be ridiculed, 124
Allopathy, a misnomer, 245, 249, 250, 251
Amenorrhœa in consumptives, 182
Analyses, why make your, at home, 203
Anæsthetics, in trifling cases, 165
    precautions regarding, 164
Are boils healthy? 188
Art of changing medicines, 174, 212
Assistant, acting as, 3, 128, 129, 221
Assistants, four excellent, 194
Attending another's practice, 103, 104, 129
    by the year, 292
Autopsies on private cases, 113
Avarice, 305
Avoid, something to, 79

Babes, hand-fed, 184, 185
Bandaging too tightly, 164
Baptism, conditional, 135
Bargains, Indian, 111
Beans, Panama, 197
Bills, 17, 280-323
    how to present, 303, 304, 315
    special color for, 322
Black lists, 322
Blame, laws that govern, 196, 197, 338
Boasting, 36, 37
Boils, are they healthy? 188
Boldness, 51
Book-agents, 85
Books, buying, 85, 86
Books, family medical guide, 180
    posting one's, 387, 390
Bores, 50, 170, 171
Borrowing, 92
Bread-pills, 151
Business system, 9
    the proper time to talk, 285, 301
Busy, too, 171, 302

Calls, hurried, 20
    list of, to prepare, 288
Cards, business, 17
Carriages and horses, 26, 27, 28
Case, withdrawing from a, 155
Cases, refusal to take, 115, 117, 171
    why never abandon, 139, 160
Cash system, the, 283
Catholic patients, duty to, 117-137
Cautions, 21, 325, 330
Censure, the laws of, 328
Certificates, death and other, 75, 125, 269, 270, 271
    clergymen's, 270, 271
    legal, 125
Changing diagnosis and prognosis, 121
    medicines, art of, 174, 212
Chapters, beginning of, 1, 35, 77, 106, 131, 163, 193, 208, 227, 259, 280, 324
Charges, increasing one's, 293
Charities, special, 310, 311, 312
Charity, the demands of, 57, 101, 309, 310
Cheap doctors, 295
Cheerfulness in the physician, 48
Children, crossness and tears in, 186
Children's influence, 54, 99
Chloral and other hypnotics, 157, 206
Chronic discharges, suppressing, 145, 188
    diseases, patients with, 138, 145
Clandestine visits, 131, 142
Clergymen, ministrations of, 131, 237, 238, 270, 271
Coincidences, good and bad, 165
Coition, why not recommend, 181
Cold rooms in sickness, 184
Collecting bills, 280, 323
Collector, 321
Commission *versus* omission, 162
Commodes in bedrooms, 187
Companions, what kind to select, 10, 11
Competition, wars of, 291, 292
Concealing presence of disease, 126, 127
Conditional baptism, 135
Conduct in the sick-room, 46-56
Confidants, 120, 122
Confidence, the, of patients, 50
Confinement, purgative after, 191
Confinements, attending women in, 114-117
Congestion, hypostatic, 114, 326, 327
Consult, right of refusal to, 214, 215
Consultation fees, 212, 213

(337)

# 338 INDEX.

Consultation, punctuality in attending, 211
  radical changes after, 212
  room, arrangement of, 6, 7, 8
  the suspense preceding a, 211
  whom to call into, 211
Consultations, management of, 121, 220
  object of, 210, 220
Consulting physician, dispensing with, 214
Consumption, errors regarding, 182, 183
Consumptives, why they cease to menstruate, 182
Contagion, fear of, 190
Contagious disease, cautions regarding attendance upon, 126, 127
Contingent fee, why not work for, 293
Contracts to do, what a physician, 70, 72, 171
Costly medicines, 274
Countenance, the physician's, 174, 324, 325, 330
Creaking boots, 92
Creed, difference between limiting one's, and limiting one's practice, 229, 230
Critics and wiseacres, 172, 173
Cuckoo, the, 32
Cures, guaranteeing, 124, 298

Dampness, 191
Death, appearances after, 326
  causes of sudden, 178, 327
Death, the power of, 132, 139, 140
Debates, how to conduct, 83, 84
Decrying medicine, 221–225
Degree of certainty in medicine, 221, 222, 227, 332
Dialogues, 112, 264
Dictionaries and encyclopædias, 37
Diet-list, 200
Dining out, 96
Discharges, suppressing chronic, 188
Discoveries, attitude of physicians toward, 230
Discussions, joint, 232
Diseases, chronic, 138
  driving in, 188
  fees in advance for attending secret, 117, 299, 300, 301
  number of mankind's, 332
  the increased tolerance of, 225, 226
  urinary, 179, 328
  venereal, 118, 300, 301
Dishonesty, where found, 312, 313, 314
Dismissal of medical attendants, 67, 68, 155, 156
Dispensary and hospital patients compared with private, 60
Dispensaries, free, 310, 311, 312
Doctor, bestowing the title of, 236
  or physician, 14

Doctoring the womb, 176
Dog-bites, 189
Donations, making, 101
Dosage, rules for, 149
Doses, heroic, 149, 152, 156
Double callings, 23, 24, 25
Dress and manners, influence of, 21, 22, 23
Dressing too warmly, 183, 184, 185
Drinking, 11, 93, 94
Druggists, 259–279
Drugs, etc., got gratis, 261
  necessity for pure, 275, 276
  that enslave, 206
Duties, five cardinal, 157
Duty to the dying, 131–138
Duty to the laws, 125, 148

Eat anything, may he? 200
Eating with patients, 96
Education, its importance to the physician, 38, 39
Emergency cases, 20, 67, 122
Engagements, making, 60, 70
Enmity, personal, 35, 36, 94
Entanglements, to avoid, 166, 167
Epidemics, 126, 127
Error, precautions to take against committing, 78, 157, 158
Eruption, driving in, 188
Eruptions, bringing out, 188
Estate of deceased physician, 318, 319, 320
Estates, charges against, 307
Ethics, medical, 60–70
Eucharist, the Holy, 135
Examinations, careless, 158
  gentleness in making, 54, 55
Examining boards, 235
Expedients, doubtful, 66
Experience, value of, 107, 108, 109
Experiments tried on patients, 159
Experts, pseudo-medical, 73, 74
Exposing false systems, 233
Extreme Unction, 134
Extremists, 238

Fainting, 189
Familiarity, undue, 10, 96, 141, 144
Family, a physician attending his own, 302
Fashionable frauds, 312, 313
Fashion in medicine, 152
Fashion and wealth, influence of, 236
Fashions, conforming to the, 22
Fear of contagion, 190
Fee in advance for secret diseases, 299, 300
  or no fee, 298
  table, 18, 291, 309
Fees, 245–323
  doubtful, rule regarding, 283, 284

## INDEX.

Fees, fixing the responsibility for, 285
  for important cases, 295, 296, 297
  how to collect, 301, 302, 303, 314, 315
  joint, 308
  lawsuits to recover, 306
  office, 18, 19
  why not work for contingent, 298, 299
Female, examining a, against her will, 165
Females, influence of, 12, 98
Feuds, professional, 35
Fever, water and ice in, 200
Fickleness, human, 130, 153–156
Finances, the physician's, 317, 321
Fingers, the tips of a physician's, 93
Foreign bodies, swallowing, 189
Forgot you, I, 117
Formulæ, private use of, 261
  stereotyped, 43
Fractures, popular error regarding, 165
Frauds, fashionable, 312, 313
Free dispensaries, 310, 311, 312
Friends, making, 47, 48, 49

German language, usefulness of the, 41
Golden rule, the, 36, 62
Gratifying whims, 168
Greek language, usefulness of, 40
Guaranteeing cures, 124, 298
Guard yourself, 151
Guides in judging whether a pharmacy is properly conducted, 268, 274
Gums, object of lancing children's, 185, 186

Habits, disgusting, 9–10, 92, 93, 96
  professional, 31, 32, 33
Hahnemann compared with Copernicus, Newton, and Jenner, 247
Hand-fed babes, 184, 185
Health, how to maintain your, 21, 101, 102, 103
  trips for, 198, 199
Heart disease, death from, 327
Hectic confounded with malarial fever, 182
Hell on earth, 175
Heroic doses, 149, 152, 156
High science, 87, 88
Holy Eucharist, the, 135
Home, not at, 171
Homœopathic creed, the, 244, 245
Homœopaths, bogus, 254, 255
  test showing which are real, 258
Homœopathy, 243–258
  is it founded on the sole natural law? 247, 248
  one of the evils of, 251, 252
Homœo versus home, 252, 253
Hope, 50
  taking away, 50, 139

Horrifying remedies, 195
Horses and carriages, 26, 27, 28
Hospital and dispensary patients compared with private, 60, 61
Hospitals, sending patients to, 199
Hours, designating the, on bottles, 206
How to conduct debates, 83
Human fickleness, 130, 153–156
  gullibility, 231
  life, value of, 55
  nature the same everywhere, 327
Humanity, its demands, 216, 217, 253, 309
Humoring the sick, 50, 168
Hurried calls, 20
Hydrophobia, 189, 190
Hypodermatic medication, 206, 207
Hypostatic congestion, 114, 326

Idiosyncrasy, 58, 241
I forgot you! 117
Important cases, fees for attending, 287, 294–297
Incompatibles, 42
Indian bargains, 116
Indorsing domestic remedies, 169
Infants, having physicians for sick, 187
Influence of dress and manners, 21, 22, 23
  of females, 12, 98
Ink, best color to use, 289
Inquiries, making, etc., in presence of strangers, 119
Instruments of precision, 20, 178
Insults, 121
Interest, evincing, in cases, 56, 57
Iron injuring the teeth, 196
Irregular physicians, contact with, 10, 215, 255
  what constitutes, 215, 217
Irregulars, joint discussions with, 232
  popular favor toward, 235, 236
  proper course toward, 10, 215, 254
  why patronized, 235–239, 252

Jealousy, 30, 31, 32, 35, 120, 167
Jenner, 256
Joint fees, 308
  discussions, 232
  practice of medicine and pharmacy, 259, 311
Junior, posing as a, 111, 112

Kindness, influence of, 50, 57

Labeling, mistakes in, 278
  prescriptions, 195, 205, 277, 278
Labels, advantage of putting the date on, 278
  putting the hours on, 206
Languages, learning foreign, 41
Latin, use of, 38, 39, 40

# INDEX.

Laws, duty to, 125, 148
    favor shown to physicians by the, 125, 126
    medical, 233, 234, 235
    their exceptional kindness to physicians, 105
Lawsuits, 72, 73, 306
Ledger, why keep a, 18
Legal duty to patients, 70, 72, 171
Liberal profession, why medicine is a, 219
Library, contents of one's, 85, 86
    creating a, 85, 86
Life-insurance, 72
    for self, 320
    power of human endurance, scale of, 332
    should a physician ever shorten? 328, 329
Limiting one's practice, 117, 229, 230
List of visits, how to prepare a, 288
Local option, ordering liquor under, 95
Locate, where to, 3–6
Longevity of physicians, 319

Malarial affections, recurrence of, 197
Malingerers, 60, 124
Malpractice cases, 72–75, 162, 163, 164
    suits, why there are more surgical than medical, 163
Mankind, study of, 44, 193, 194, 331
Mankind's dependence on physicians, 333, 334
Manners, influence of, 37, 47–49, 51–54, 92, 93, 99
Marriage of physicians, 97
    of syphilitics, 181
Marriages, unlucky, 97, 98
Maxim, 312
Medical profession, greatness of the, 333, 334, 335
    art, imperfection of, 221, 222, 223, 227
    ethics, 61–70
    examining boards, 235
    societies, 81–84
Medicine, decrying, 221, 222, 223
    degree of certainty in, 221, 222, 223, 227
    fashion in, 152, 240, 243
    fear of, 193, 194, 195
    haters, 193
Medicines, art of changing, 174
    at office, 19
    bad effects of, 195
    charges for, 262, 263, 274
    costly, suggestions regarding, 263
    palatability of, 193, 194
    the dynamization of, 248, 249
    unused, 174
Memory of cases, 85, 145
Menial labors, 122

Mental therapeutics, 150, 151, 194
Metric system, 90
Microscope, working with the, 114
Midwives, assisting, 116
Milwaukee physicians, offer of the, 248
Mineral *versus* vegetable medicines, 196
Mischief-makers, 120
Mistakes of pharmacists, 266, 277, 278
    in compounding, cases of, 266
Moralizing, 129
Morphia granules, 242
Motto on bills, 312

Name, what is in a? 249, 250, 251, 253
Neglectful and perverse patients, 168
Neighborly visits, 30
New remedies, how aided, 194
Newspaper squibs, 29, 103, 170
Night emissions, 182
    visits, 142
No cure, no pay, 298
Nostrums, why condemn, 273
Novelty, influence of, in medicine, 168
Number of mankind's diseases, 332
Nurses, conduct toward, 167, 186

Objects of consultations, 210, 220
Obstetrical cases, 114–117
Office, absence from, 15
Office, charges for advice at, 18, 180, 297
    hours, 16, 17, 18
    location and arrangement of, 5–9
    outfit, 6, 7, 19
    practice, 16–19
    signs, 14–18
    students, 12, 13
Offices, branch, 5
Old lady with salve, 170
    persons, operations upon, 328
Old woman of Paris, 151
Omen, a bad, 137
Omission *versus* commission, 163
Only sprained, 330
Only drunk, 330
Opiates, their place and power, 156, 157
Opinions, necessity for caution in giving, 53, 123, 324, 325, 329, 330
    of patients and their attendants to be considered, 186
    that terrify, 175–178
Other physicians, attending after, 64, 65, 103, 104, 308
Overlooking diseases, 157, 158
Overpraise from patients, 119, 120
    from relatives, 120
Overvisiting, 122, 143, 144

Panama beans, 197
Partisan questions, 26, 94, 270
Partnership, 2
Passions, influence of, 147, 148

# INDEX. 341

Patients, dining with, 96
  distant, 117
  foreigners as, 41
  hospital and dispensary, compared with private, 60, 61
  how to transfer, 209
  physician's legal duty to, 70, 72, 171
  neglectful and perverse, 168
  purse-proud, 154
  quoting authorities to, 111, 112
  refusal to take, 171, 172, 313, 314
Patients, rights of, to additional advice, 214
  varieties of, 141, 193, 194, 331, 333
  whims gratified, 50, 168, 186, 239, 240, 242
  worthless, 146, 313, 314
Pay you, when shall I? 287
Paying one's debts, 91
Peculiarity of manner, 52
Pencil sketches, 160
Penmanship, 41
Pension claimants, certificates to, 75
Percentage from pharmacists, 260, 261
Personal affairs, privacy regarding your own, 321
  appearances, 21, 22, 23
Pharmaceutical catch-pennies, 204, 205
Pharmacists, 259-279
  indiscreet, 266
  prescribing, 265, 311
Pharmacopœia, breadth of the U. S., 272, 273
Phlegm, swallowing, 186
Photograph giving, 130
Physician, a sickly, 45
  estate of deceased, 318, 319
  how an irregular may become a regular, 219
  mission of the, 140, 214, 309, 333, 334, 335
  or doctor as a title, 14
  the new, 30-33
  *versus* doctor as a title, 14
  what constitutes a regular, 217, 219
Physician's countenance, the, 174, 324, 325, 330
Physicians, assaults upon, 165
  associates of, 10, 11
  drunken, 11, 93, 94
  increase of number of, 34, 35
  sickly, 45
  taking office, 23, 24, 25
  the marriage of, 97
  why the older excel the younger, 107-109
  why they do not get rich, 316-321
  young, 4
Placebos, 150, 151
Pocket visiting-list, 18
Policy *versus* Principle, 168, 243

Politeness, value of, 98, 99, 100
Politics, 9, 25
Pollution, self, 182
Polypharmacy, 202
Poor, attending the, 57, 58, 59, 309, 312, 322
Popular test of skill, 49, 86, 88
Posting one's books, 287, 290
Post-mortem discolorations, popular error regarding, 326
Post-mortems and analyses, 113, 114, 326
Practice, difference between limiting one's, and limiting one's creed, 230
  limiting amount of, 117
Practice, preparing one's self for, 2, 45
  soliciting, 153
Precautions, 157, 164
Pregnancy suspected, 78, 79, 158
Prescribing, extravagance in, 174, 175
  suggestions on the subject of, 156, 157
  without an interview, 221
Prescription, object of a, 262
  papers, 260
  the, to whom does it belong? 262
Prescriptions, about labeling, 205, 206, 277, 278
  expertness in writing, 42, 43, 191, 202
  joint, 206
  ready written, 111
  unauthorized renewal of, 262
Presents from patients, 95
Press notices, 104
  from physicians, 104
  writing for the medical, 88-91
Principle *versus* Policy, 168, 243
Private formulæ, use of, 261
Production of abortion, 77, 78, 79
Profession, identifying self with the, 80, 81
Professorship, choice of a, 25
Prognosis, cautions concerning, 53, 110, 111, 121, 138, 139, 160, 325, 330
Provings, homœopathic, 252
Purgative after confinement, 191
Purity of mind, 46

Quack bitters, 130
  medicines, their proper position in the apothecary-store, 277
  relation of pharmacists to, 265, 274
Quackery, ours the age of, 233
Quackish methods, 28
Quacks and impostors, depredations by, 172, 180, 233
Quantities to prescribe, 174
Question, an awkward, 111
Questions, asking private, 119
  rule for repeating, 192
  rule regarding, 47, 158

Questions, unwelcome, how to avoid, 139
Quinine, popular prejudice against, 196
Quoting what the books say, 111, 112
Quotations at beginning of chapters, 1, 34, 77, 106, 131, 163, 193, 208, 227, 259, 280, 324

R as a sign or symbol, 43
Receipts, why compel people to take, 305
Reclaiming those who have strayed, 232, 233
Recommending other physicians, 172
Record-book, 159
Recreation, necessity of, 101, 102, 103
Reference-book, 159
Register heat, 184
Religion, 131-141
Relinquishing attendance, 142, 212, 214
Remedies, domestic, indorsing, 169
    examined at every visit, 202
    horrifying, 195
Removals, frequent, 5
Renewal of prescriptions, to prevent, 205
Reporting cases, 88, 89
Reputation, value of, 44, 45
    varieties of, 29, 44
Responsibility, dividing the, 162, 163
Rich not to pay for poor, 294
    the poor who get, 154
Riding *versus* walking, 26
Rivalry, professional, 31, 32, 35
Routine practice, 43
Rule for repeating questions, 192
    the golden, 36, 62
Ruling spirits in family, the, 120

Scandal, 36, 46, 47, 157
Scarlet rash, 187
Scold, how to, without offending, 97
Secrets, 119, 146, 147, 148
Selection of a location, 3-6
Self-medication, 203, 204
Self-pollution, 182
Self-preservation, 30, 31
Self-reliance, its value, 208
Seniors, respects due our, 106-109
Sequelæ, foreseeing, 330
Servants, attending, 59
Services for employers, 285
    to clergymen, 271, 307
    to physicians, 307
    to the poor, 57, 58, 309, 310, 322
    when not to charge for, 306, 307
Sexual intercourse, why never recommend, 181
Shame, a matter of, 166
Short visits, how to make, 122, 200
Shut your eyes, to these, 187
Sick, humoring the, 50
Sickly physicians, 45
Signatures, regarding, 251
Signs, office, 14, 15

Skill, what is medical, 43, 56, 222, 223, 332, 333
Slavery of a physician's life, 101, 102, 103, 315-320
Small-pox, popular error regarding the, 71
Social influence, 100
Society, duty to, 148
Soliciting practice, 155
Something to avoid, 79
Speaking-tube, 17
Specialties, when to patronize, 208, 209
Specialty, adoption of a, 59
Specifying the particular make, 273
Speculum, abuse of the vaginal, 177, 179
Spoons, variations in size of, 206
Sprain, fracture or dislocation, 330
Spring-time, taking medicine in the, 152
Stepping-stones to practice, 57, 58, 59
Streets, barricading or roping the, 192
Students, increase of, 13, 34
Suffixes, adding to one's name, 91
Sunday as a day of rest, 102, 279, 287
    prescribing liquor on, 95
    practicing on, 102, 103
Superseding other physicians, 64, 65, 103, 104, 308
Suppressing chronic discharges, 188
Surgeons, why demand for, is limited, 51
Surprise, showing, 324, 325
Syphilitic cases, 175, 299, 300, 301
Syphilitics, the marriage of, 181
System in business, 9
Swallowing foreign bodies, 180
    phlegm, 186
Swearing off from drink, 129
Sweating during sleep, 200
    the baby, 185
Synonyms, use of, 204, 276, 277

Telephone, the, 17
Temper, control of, 53, 101
Terms, transposing, 277
Terrifying opinions, 175-178
Therapeutics, crude, 193, 194
    mental, 150, 151, 193, 194
Thermometer, clinical, 21
Time, rapid flight of a physician's, 101
Time lost in waiting, 201
Toleration of difference of opinion, 84
Tongue-depressors, 128
Trade-mark articles, 204, 272, 273
Transposing terms, 277
Tricks, 28, 33, 121
Trips for health, 198, 199
Triumphs, 65
Truth, its value, 138, 249

Unction, extreme, 134
Urinary diseases, 179, 328
Urine, scanty and high colored, 189

# INDEX. 343

Vacation, the physician's, 101, 102, 103
Vaccination, 70, 71
Value of politeness, 98, 99, 100
    of reputation, 43, 44
Variability of human endurance, 332
Variation in the size of spoons, 206
Varieties of patients, 193, 194, 331
Vegetable *versus* mineral medicines, 196
Veins on back of hands, 191
Venereal cases, 117, 118, 299, 300, 301
Visit, conduct at, 46–56
    extra charge for the first, 294
Visiting-list, best way to carry one's, 289
    how to improve a, 288
Visiting the patient of another physician, 64, 65, 103, 104, 129
Visitors to the sick, 191
Visits, how to make short, 122, 200
    neighborly, 31, 80
    to the sick, 49, 120, 142, 144, 191, 192
Vivisection, 114
Vocabulary of different classes, 331
Volunteer services, 71

Waiting, time lost in, 201
Walking *versus* riding, 26, 27
Warming the newborn, 184, 185
Wars of competition, 292
What is in a name? 249, 250, 251, 253
Whims, gratifying the, of patients, 50, 168
Why charge for every visit, 295
Wife, the meddling, 146
Wills, 126, 137
Wiseacres and critics, 173, 174, 203
Withdrawal from a case, 155
Witness, duty to self when a, 73
Woman, devotion of, to the sick, 56
Womb-doctoring, 176
Work, amount of, done by every busy physician, 315–319
Worms, has he? 186
Worthless patients, 171
    systems of practice, how to expose, 233
Writing for the medical press, 88–91

Youthful physicians, 4, 111, 112

www.ingramcontent.com/pod-product-compliance
Lightning Source LLC
Chambersburg PA
CBHW032355230426
43672CB00007B/712